# THE
# INTERNET
# REVOLUTION

## The Not-For-Dummies Guide to the History, Technology, and Use of the Internet

## J. R. Okin

**Ironbound Press**
**Winter Harbor, Maine**

First Ironbound Press edition, 2005

*The Internet Revolution: The Not-For-Dummies Guide to the History Technology, and Use of the Internet* © 2004 by J. R. Okin All rights reserved. No part of this book may be reproduced in any form by any electronic or mechanical means (including photocopying, recording, or information storage and retrieval) without permission in writing from the publisher. For information, address: Ironbound Press, P.O. Box 250, Winter Harbor, ME 04693-0250.

Ironbound Press books may be purchased for educational, business, or sales promotional use. For information, please write: Special Markets Department, Ironbound Press, P.O. Box 250, Winter Harbor, ME 04693-0250

Cover image, *The Blue Marble: Land Surface, Ocean Color, Sea Ice and Clouds*, courtesy of NASA's Visible Earth, located at: http://visibleearth.nasa.gov/

Ironbound Press Web Site: http://www.IronboundPress.com
Ironbound Press email inquiries: info@IronboundPress.com

Printed and bound in the United States of America.
Printed on acid-free paper.

Library of Congress Control Number: 2004116990

ISBN 0-9763857-6-7 (cloth)
ISBN 0-9763857-5-9 (paper)

10 8 6 4 2 ✻ 1 3 5 7 9

*To my mother and father,*
*Vera and Eugene,*
*for encouraging me,*
*for punishing me,*
*for loving me,*
*for teaching me to appreciate the world's beauty,*
*diversity,*
*and magic,*
*and,*
*not least of all,*
*for trying to have a girl after having two boys,*
*and not being disappointed.*

Other books available from Ironbound Press by J. R. Okin:

- The Information Revolution: The Not-For-Dummies Guide to the History, Technology, and Use of the World Wide Web

- The Technology Revolution: The Not-For-Dummies Guide to the Impact, Perils, and Promise of the Internet

# Foreword

It's startling to think how different our world would be without the Internet. Most of us take the existence of the Internet for granted. We started using it one day — to send an email, to search for some information, to play a game, or to buy a product — and then we began to use it more and more, until, eventually, we came to rely on it. Most of us have stopped, at one time or another, to wonder what the Internet is, how it works, or where it came from. But none of these questions ever led us to reduce our use of or diminish our enthusiasm for the Internet. That's because we didn't simply welcome the Internet into our lives, we embraced it; and, in many ways, we changed how we live as a result.

Yet the vast majority of us who routinely use the Internet today had never sent an email or heard of the Web just ten short years ago. And ten years before that, in the early to mid 1980s, when only a very small number of us worked with computers (and the majority of computers were not yet attached to a network) few of us had heard of the Internet. How did the Internet become such a commonplace and integral presence in the lives of so many people in such a short period of time? This was the first question I asked when I began the research that led me to write this book.

Think of the Internet as a new and revolutionary transportation system that carries one of the principal commodities of our time: information. This transportation system has become as fundamental to the daily functioning of our world as the network of roads that interconnects our cities, towns, and villages. Like the roads, the Internet accommodates a variety of vehicles that are used to carry its traffic. These vehicles transport an endless variety of packages, and they deliver them with a speed and efficiency unrivaled by any other transportation system. And, in its own way, the Internet even accommodates us as passengers, carrying information about us at speeds approaching the speed of light.

No matter what perspective, previous experience, or preconceptions we bring with us when we use the Internet, whether we are novices with little or no understanding of its technology or experienced professionals working in the computer field, it is hard not to marvel at how well the Internet works and at how much of our world and of ourselves has been changed by it. My introduction to the Internet came in 1985 and, while its significance in my life has changed over the years as I have moved from job to job and as the Internet itself has evolved and expanded its reach into our homes, places of work, libraries, and elsewhere, my fascination with the graceful simplicity of its operation and the unlimited potential of its use has never abated.

In 1985, shortly after the small computer software company that I was working for in New York City lost its financial backing and was shut down, I was hired as a consultant to AT&T Information Systems in Freehold, New Jersey, to manage documentation projects for its technical publications department. I arrived at AT&T with some marginal computer skills and a sound knowledge of and appreciation for English literature, thanks to six years as an English major, first at Vassar College and then at Cambridge University. The technical publications department of a large telecommunications company seemed like a good place to pay off some student loans, stay connected with books, and broaden my knowledge of computers. I never would have guessed that it would also prove an ideal place to explore the relatively new area of computer networking and the still small, but highly dynamic, environment of the Internet.

That first consulting job with AT&T marked the beginning of nearly twenty years of consulting (and continual Internet access), as I took on work as a programmer, tester, systems engineer, and manager for such companies as AT&T Bell Laboratories, Lucent Technologies, Unix International, and the Open Software Foundation. When the dot com boom arrived in the late 1990s, I returned to New York City to work as the director of software engineering and development at a small Internet startup called Linkshare. About two years later, when the dot com bubble burst, I quickly (and eagerly) left Silicon Alley (the area in Manhattan near Union Square where Linkshare was located and that was so named because of its high concentration of Internet startups) far behind me and moved my life to the rocky coast of Maine, where the idea for this book was born and took shape.

I consider myself fortunate to have enjoyed access to the Internet in its earlier form as an experimental network used largely by academic and research organizations. I didn't know much about its technology back then, but I quickly discovered ways in which to apply its capabilities in the performance of my work, as many people routinely do today. I saw firsthand the Internet's transformation from a publicly funded, U.S.-controlled research project to a privately funded and privately controlled international network. I marveled at the introduction of the first Web browsers and, along with many other programmers, I helped to create some of the earliest Web sites on the Internet. I watched as the Internet gradually changed into a new type of information resource; the effect was like that of searchlights coming on the darkness, one after another, each one illuminating a new library of information that housed a unique collection of books, periodicals, programs, and data. I also watched as the Internet changed again with the arrival of big business, the introduction of commerce, and the slow but unavoidable additions of advertising and commercialism. Finally, I put myself at the center of the dot com maelstrom when I joined Linkshare in 1999 and set out to help engineer, develop, and manage the company's technology using the proven practices, processes, and techniques of established companies like AT&T and Lucent Technologies. While at Linkshare, I witnessed the greed and unrealistic expectations that were common during the dot com boom and the very personal and tragic consequences of the dot com bust.

My experiences with the Internet — as an ordinary user, as a systems engineer and developer, and as a dot com manager — and my fascination with the impact of the Internet and its technology inspired me to write a book about the Internet. My goal was to combine the story of the Internet's creation and development with an explanation of its technology that anyone could understand and an exploration of the Internet's impact on our lives, jobs, and community and social structures. After two years of research and writing, I created a book that presented all of these subjects and that was divided into three nearly equal parts.

Upon reviewing the completed book, I realized that its size and scope might alienate the audience I most wanted to reach: individuals who wanted to understand how the Internet worked and why it had succeeded in becoming such a revolutionary presence in our lives. So I decided to publish each of the three parts as a separate book. The present volume, which focuses on the history and technology of the Internet, is the first. The second book, which focuses on the history and technology of the Web, is entitled "The Information Revolution." The third book, which focuses on the impact of the Internet's and the Web's technology, is entitled "The Technology Revolution." I believe that each of the books can be read and understood (and, hopefully, enjoyed) independently of the other two. Together, the three books present a broad and thorough depiction of the revolution of our times, that began as a remarkable technological innovation but became — because of us — a force that has changed and is changing the way we work, play, ask questions, find answers, buy and sell products, communicate, and interact with others.

J. R. Okin
December, 2004

# Acknowledgments

Writing is a solitary and sometimes lonely pursuit. But producing a book is a collective effort that owes its completion to many different types of contributions from a large number of individuals.

To my family, who, over the last three years, expressed support, urged me on, sheepishly asked how I was doing, and eventually were kind enough to stop inquiring about the book, I convey my thanks (for all of the above). More specifically, to my father, to Peter, Lisa, Daniel, and Stephen, to Rick, Thea, Sara, Emily, and Megan, thank you all for your patience and for the gentle nudges conveyed in words said, and left unsaid, to finish up the book and move on. I must convey a special thanks to Toni DeAngelis, whose nudges resemble kicks and whose words of encouragement often come in the form of marching orders, but whose caring and love are nonetheless transparent and are always overwhelming.

To my 'extended' family in Maine, I owe a separate measure of thanks. The questions, kind words, and encouragement I received from Mary and Greg Domareki, their children, Sarah, Catherine, John, Greg, Luke, and Bridget, Sarah's husband, Mike Kazmierczak, and their son, Michael, and Greg's wife, Erin, and the rest of their large and loving family helped me in no uncertain terms to stay the course and find my way to the completion of this project.

To my friends, who answered questions, provided solicited (and unsolicited) advice, checked in on me, and reminded me again and again that there was an audience for the book, I am forever grateful. I owe thanks to Carol Schur, who was kind enough to read an early draft of the book and provide valuable feedback, to Ann Hagerman, who was always generous with her suggestions and opinions, to Barry Orr, who called and emailed and was never in short supply of humor or kind words, to Helene Armitage, who offered support, suggestions, and encouragement when I first started out, and, off an on, throughout the course of the project, and to Ellen Dreyer, who provided much needed information about the inner workings of the world of publishing along with some gentle words of encouragement.

To my editor and good friend, Bob Lippman, who never tired in his efforts to correct my mistakes, to question my arguments and conclusions, and to identify material that needed further explanation or that demanded simplification, I am in your debt. I could not have asked for a more skillful or thorough editing of the book; and, thanks to you, I now have a new appreciation for the meaning of the phrase, brutal honesty. I am grateful that you found the time to take on this project and I very much appreciate your commitment to seeing the work through to its conclusion. Whatever value or use this book may find, it was greatly enhanced by the time, effort, and skill you put into editing its contents.

Finally, to Mary Domareki, who kept me from giving up, who held my hand, who cajoled me, and who never wavered in believing in me, words are inadequate to express how I feel. I will have to work hard to find a way to repay you. But I will find a way.

# Contents

# Contents

## 4  Packet Switching: Lifeblood of the Internet

## 5  Protocols: The Definition of Interoperability

# Contents

# Contents

## A Milestones, Netiquette, and Jargon

## B Notes

## I Index

# The Internet Revolution

## Why a Revolution?

Not very long ago, the only people talking about the Internet were the small number of individuals who were engaged in engineering it, building it, and nurturing its growth. That was during the 1970s. But even during the 1980s, as the Internet expanded its reach and diversified its information resources and services, the Internet's existence went largely unnoticed by the general public and it managed to retain its quiet, remote, and unassuming presence. The only national press coverage the Internet received during the first two decades of its existence was when there was a sudden and sizable failure in one of its component systems or a debilitating network attack, like the Internet worm of 1988. And the only individuals who took an interest in the Internet were computer enthusiasts, and their numbers were still very small. Moreover, in order to have any access to the Internet, you had to work at a major research facility, like AT&T Bell Laboratories, or attend one of the lucky few Internet-connected academic institutions, like MIT or Stanford, or be employed by one of the United States government agencies that controlled it. In other words, for the first twenty or so years of its existence, the Internet remained predominantly hidden from public view and functioned as the private and entirely commerce-free playground of researchers and scientists, students and computer center workers, and some members of the military.

Nowadays, it's difficult to avoid some reference to the Internet, no matter how hard we may try. It comes up in conversations with friends and family and in meetings at work as people complain about the latest email virus or some interminable network slowdown or as they rave about a site they have just discovered for managing their stock portfolio or playing poker. We hear about the Internet on radio and television programs as broadcasters awkwardly spell out their Internet site addresses in an effort to entice us to get more in-depth information on a particular subject, such as a recent news story or the upcoming episodes of a popular show. We see the Internet's effect on commerce everywhere we look, in its role as an entirely new and powerful medium for the distribution of all forms of advertising, as evidenced by the Web site names that confront us on everything and anything that can contain printed text, from billboards to baseball caps, t-shirts to tattoos, the sides of cars, trucks, trains, and planes, the walls of sports stadiums, the cellophane wrappers enclosing heads of lettuce, and even those annoying stickers attached to each and every grapefruit, tomato, and cucumber.

This change didn't happen overnight, but it definitely feels like it did. It took less than ten years after the Internet was privatized and opened up to commercial traffic in the early 1990s for it to become a basic and essential part of our daily lives. It became, quietly and unobtrusively, an integral component of our home life, our jobs, and the world's communication infrastructures, economies, and cultures. This change was so compelling and pervasive that it raises the question: what kind of vacuum or void existed before the Internet's creation and its evolution into such a powerful and ubiquitous presence? In the course of a few, frenetic years, the Internet has grown into an inexorable force that businesses, non-commercial organizations, governments, scientific and academic institutions, and individuals throughout the industrialized world have not only accepted but embraced. Consequently, the Internet's impact can be seen all around us.

The Internet has transformed how business is conducted, and it has provided powerful new ways to locate, learn about, and buy all types of products and services. It has inspired and made possible the creation of entirely new business enterprises, including the much touted and highly speculative business of e-commerce. It has enabled governments to better share information about and distribute information to their citizens, and better collect information about those citizens. It has facilitated collaboration on research, which, incidentally, fulfilled one of the visions of its original creators. It has dramatically changed the way we communicate and has enabled the creation of new social structures in the form of virtual communities. It has forever altered how we access information and the variety and quantity of information we can access, empowering us to gain knowledge through a richness of resources that was previously only imagined in science fiction. It has allowed us to become publishers of family photos, shared genealogies, journals, diaries, diatribes, musical compositions, short stories, full length novels, and just about anything else that can be stored and distributed in the form of a computer file.

For the purposes of a simple comparison, consider an earlier revolution incited by a different, but just as pronounced, leap in technology. The Industrial Revolution, which began in England in the 18th century and resulted from the invention and refinement of the steam engine, brought us the mechanization of labor. Machines were created that performed the labor of men, women, and farm animals, and they performed this work faster, cheaper, with fewer problems and interruptions, and often with greater precision. For a fortunate minority, the Industrial Revolution meant new-found wealth, provided one understood how to apply the new technology and succeeded in financing and managing a new type of business venture. For the majority of the population, it meant a change in employment and, more often than not, a resulting change — for better or worse — in one's financial situation, as many old, established jobs were eliminated or greatly changed and new, very different jobs were created.

The impact of the Industrial Revolution and the mechanization of labor was immediately evident in changes in the job market, the creation of new businesses and new products, and the quality and quantity of products that could be produced. But its most profound and lasting impact — albeit less immediately apparent — was revealed in how individuals lived their lives and interacted with others. It not only affected what people did for employment, it affected how people performed their jobs. It also affected people's home life and the amenities in their homes, their health and the general comfort and quality of their lives, their opportunities for education and advancement, and how fast and how far they (and their information) traveled the world.

In some way, shape, or form, the Industrial Revolution eventually affected every region of the world and nearly every member of society. Even the few, isolated areas of the world that, for whatever reason, failed to feel its direct impact, were eventually indirectly affected either through the trade and transportation of goods, the communication of information, the expansion of urban areas into rural communities, or the increased movement and migration of people. Today, more than two centuries later, some segments of society remain distant or disconnected from the Industrial Revolution's mechanization and associated modernization, a few through choice (e.g., the Amish) and others due to lack of financial resources or other limiting economic factors. Even so, whether people rejoiced in its arrival, felt indifferent to it, shunned its existence, or somehow avoided its impact, the Industrial Revolution changed forever the face and form of the world, and, directly or indirectly, these changes had consequences for everyone.

The Internet Revolution, which began in the U.S. in the early 1990s and resulted from the proliferation and internetworking of computers, is reshaping our world right now, whether or not we are a willing and eager participant and whether or not we want to acknowledge it. It has brought about a different sort of mechanization than that brought about by the Industrial Revolution, but one equally broad and far-reaching in its impact: the mechanization of information and communication. Moreover, there is no going back, no undoing of its effects, any more than one could undo the effects of the Industrial Revolution. Also, much as the Industrial Revolution changed the lives of individuals in different ways, the Internet and its revolution in how we communicate, acquire information and educate ourselves, perform our jobs, entertain ourselves, contribute to our communities, and interact with others, means something different to each of us.

If you currently use the Internet, consider for a moment how much time you spend online at home, at work, or elsewhere. Think about the information you routinely access through the Internet or the amount of email you send and receive. Two of the most popular Internet services — email and the Web — are used by millions of people across the globe each and every day. These services constitute only a small fraction of those the Internet offers. But they alone have changed the way we interact with our friends, family, and others, the variety and volume of information at our disposal, and, more generally, how we conduct our lives.

The Industrial Revolution was not a revolution because the mechanization it brought changed the way products were manufactured. Nor was it a revolution because it enabled the creation of entirely new types of products and services or new sorts of jobs. Although it did all these things and the effects were both permanent and far-reaching, what made it a revolution was how it transformed us. The same is true of the Internet Revolution.

Everywhere we look, we see more and more references to the Internet. That's because it is becoming part and parcel of everything we do. The Internet is changing how we raise and educate our children, how we stay connected with our families and friends, how, when, and where we perform our jobs, how we purchase our goods, how we read the weather forecast or our horoscope or send a birthday card. These changes in our behavior are fundamental and permanent, and they are becoming more pervasive with each passing year. Consequently, the Internet is changing us, our communities, our societies, and, as you will understand after reading this book, the very interconnectedness of our world.

## What is the Internet?

The very short definition of the Internet is that it is a network of computer networks. As such, the Internet comprises a communication infrastructure that enables computers to locate and talk to one another and to send and receive information. The Internet's infrastructure consists of a global communication network of copper telephone lines, fiber optic cables, coaxial cables, and satellite systems. Its communication infrastructure for handling the transmission of information can be compared to the roadway infrastructure for handling the transportation of vehicles. Computers use the Internet's global network to move small packets of data quickly and efficiently from place to place (i.e., from computer to computer) and deliver them to their intended destination (i.e., another computer) much as cars and trucks use the network of roads to move people and packages from place to place. Moreover, the speed and efficiency of the Internet's system for moving information — critical factors contributing to its use and growth, as you will see — and the overall functioning of the system depend entirely on strict adherence to certain common, general rules and to other more localized regulations, much as the operation of a car or truck depends on adherence to general laws on how to drive and to local regulations relating to each particular roadway. These Internet rules and regulations are described, defined, and implemented by something called protocols.

As will be made clear in later chapters, protocols literally define how the Internet works and, by extension, what constitutes the Internet. Protocols are responsible for specifying how data moves from computer to computer across the Internet, and they work the same way whether the data's journey consists of traveling across town or traveling across the globe. Protocols are also responsible for specifying how your computer locates other computers on the Internet, how it connects to other computers, and how it arranges for the safe and reliable exchange of information between itself and other computers. Separate protocols exist to define how a file is downloaded (i.e., copied from another computer on the Internet to your computer), how a page on the World Wide Web is viewed, how messages in a chat room are sent and received, and how countless other activities are handled. Moreover, protocols are responsible for ensuring that any and all types of computers are welcome and equal on the Internet, regardless of size, manufacturer, operating system, and any number of other factors. Protocols are, therefore, what enable all computers to talk to one another; these computers include everything from handheld devices (e.g., cell phones and personal digital assistants or PDAs) to laptop and desktop personal computers to mainframes and supercomputers, from Apples to IBMs to Dells to Suns, from MacOS to Windows XP to Unix and Linux. It is this fundamental interconnectedness that best defines both the composition of the Internet and the philosophy and engineering behind its creation and development.

Today's Internet has benefited from the contributions of countless individuals, groups, organizations, and corporations. It continues to benefit from such contributions and to change because of them. This explains, at least in part, why the Internet is such a highly dynamic and fluid environment; it is changing all the time. Even a few short years from now, the Internet may well function differently and look quite different than it does today. That's because its technology continues to evolve, just as we continue to make greater use of it and discover new ways to integrate it into our lives. Furthermore, every day we help to recreate the Internet. We shape it because we embody it (i.e., our presence is on it) in the form of the information we create and request, the information that is captured about us, the identities we assume through creating accounts and acquiring usernames and aliases, and the computers we use to connect to the Internet.

Every time we access the Internet, our computer becomes part of its network, extending the Internet and further defining it by becoming another interconnected device capable of sending and receiving data across its global communication infrastructure. Like our computers, we also take up residence on the Internet; we do so, as you will see, in numerous subtle and profound ways. That's because the Internet is first and foremost a network of people. We sit in front of the millions of computers that are part of the Internet's global network; and the information that courses through its endless wires and cables is, overtly and implicitly, all about us, by us, and for us.

## Why Now?

The Industrial Revolution traces its beginnings to the invention of the steam engine. The subsequent refinement and the imaginative application of the steam engine's new technology transformed that invention from a marvel of engineering into an engine of change. The Internet Revolution is also rooted in the creation of new technology. It traces its beginnings to the invention of computer networking, which can itself be traced to another significant technological breakthrough: the launching of the world's first satellite, Sputnik I, by the U.S.S.R. in 1957.

The story of the Internet is, in large part, one of technological advancements; its advancements, moreover, are closely tied to the evolution of computer technology. The Internet's story is full of both the frequent, incremental developments in technology that continually improve on what has come before, and the far less frequent, but more momentous, leaps in technology that seem to instantaneously rewrite the rules regarding what can and cannot be done. The Internet Revolution, however, is a story about us and our times. It begins with and is inextricably tied to the technology of networking and of computers, but its focus is on our acceptance and use of the technology today. The Internet Revolution is the result of our transforming the technology into an engine of change.

The Internet owes its existence to the invention of networking and packet switching (a methodology that allows computers to exchange information efficiently and economically) in the 1960s, to the engineering and adoption of TCP/IP (the protocols that reside on all Internet-connected computers and that are responsible for ensuring that all computers can intercommunicate regardless of their size, type, manufacturer, operating system, etc.) in the 1970s, and to the development of the personal computer and its widespread acceptance, at home and at work, in the 1980s. It owes much of its popularity to the creation of the World Wide Web in the early 1990s and to the Web's effect of accelerating the integration of commerce with the Internet. This, in brief, explains the technological factors that were responsible for the Internet becoming such a pervasive, and even invasive, presence.

Technology alone, however, does not, bring about change. Neither steam engines nor computer networks could have impacted the fabric of society without factors outside their respective technologies also playing a part. This is particularly true of the Internet, which operated for more than two decades before it began to have any substantial impact on us, our lives, or the world in general. In addition to the advances in technology, it took the fortunate coinciding of several independent events in the late 1980s and early 1990s to bring about the creation of today's Internet and to incite its subsequent exponential growth.

One event was the privatization of the Internet's infrastructure, which allowed migrating the control and flow of traffic from systems operated and/or funded by the U.S. government to comparable systems operated and owned by private organizations. Another was the explicit removal of the prohibition against any commercial use of the Internet, which allowed commercial traffic to start flowing freely across the Internet. Another was the U.S. government's mandated adoption of TCP/IP shortly before its privatization of the Internet, which contributed significantly to the Internet's fundamental interoperability. And another was the introduction of information management tools on the Internet, which were responsible for making the Internet's remote and distributed resources findable and usable. The best known of these information management tools was the World Wide Web, which began operating on the Internet in 1990 (with the introduction of the world's first — and for a brief time only — Web site).

We started to arrive on the Internet in large numbers soon after all these events occurred. The Internet's size and influence, measured in the number of connected computers, the number of users, and the number of services it offers, has grown exponentially ever since.

## Why a Book on the Internet?

Because the Internet is here to stay. Because we are relying more on the Internet every day, personally and professionally, to assist us with running our lives and running our businesses. Because very few people know anything about the Internet's history, technology, uses, or impact.

The purpose of this book is to provide a working definition and exploration of the Internet in plain, non-technical language: to describe what constitutes the Internet, how the Internet works, how it evolved, the information it contains and controls, its impact on our day-to-day personal and professional lives, and why it has succeeded in becoming such a powerful presence in our lives.

Contrary to popular misconceptions, the Internet is not an intangible entity that defies any kind of clear and concise description or non-technical explanation. In fact, its form and substance, as well as the way it operates, has been described and defined in excruciating detail; this was a necessary prerequisite to its creation. In general terms, the Internet comprises a collection of relatively small and modular, cleverly engineered (i.e., refined to be as simple as possible) components that work together for the express purpose of enabling different types of computers, in locations near and far, to talk to one another, transmit data, and share information. Its purposely large-scale and generalized design and engineering necessarily makes the Internet a large subject to cover. But, by examining each of the Internet's different components and how they interoperate, you will come to understand exactly what the Internet consists of and how we interact with it. What may, however, be the most remarkable thing about the Internet is how its engineering functions to make its large, ever expanding, seemingly nebulous presence into something simple, usable, approachable, and malleable to each person's needs. Describing how this is done is one of the main objectives of this book.

Computers on the Internet accomplish their intercommunication, or networking, much as we interact and communicate with other individuals. They begin the process of networking by following some very basic, predefined rules to open a dialogue; they look up the address for, locate, and connect to another computer, much as we look up an individual's name in a telephone book in order to call him or her or locate an individual's street address on a map in order to drive to his or her home. They start up a conversation by exchanging greetings and establishing a fundamental framework about how they will communicate with each other, much as we introduce ourselves and overtly or implicitly indicate that we speak French or English or that we want to discuss last night's football game or a specific book or movie. They then proceed to perform the primary function of networking, that of exchanging information (e.g., forwarding the contents of an email message or requesting and then receiving the contents of a Web page), much as we share information in our conversations.

In simple, conceptual terms, this is precisely how the Internet functions. The technical details of how the Internet's individual components operate to accomplish this, how these components work together, and how these components were originally designed and engineered, are equally understandable. Moreover, when these details are presented outside of their specific implementation on any one computer, they make for meaningful and compelling reading that anyone can understand. As you will see, computers are, to a large extent, modeled to function in ways that are strikingly similar to how we behave. They understand and accept (and even reject) instructions, and they carry out or delegate accepted instructions as needed and as appropriate given other existing conditions and circumstances. They interpret, store, and recall information. They communicate and share information across a network. In other words, you do not have to know how to program a computer to understand how a computer — or a computer network for that matter — functions, or how to put a computer, or a computer network, to work for you. Nor do you have to be treated as a dummy and avoid the details of how the technology works and how it was created in order to feel more at ease in exploring the subject.

The story behind the creation, development, and evolution of the Internet is as entertaining and engaging as it is educational. That's because it is, first and foremost, a story about people: people facing and overcoming technical challenges to build the Internet and people employing the Internet, in unanticipated and ingenious ways, to create new, enriching, and empowering services with far-reaching consequences, such as email and the World Wide Web. Learning how the Internet developed and how its various services work should even help you use the Internet with more confidence and greater personal security.

If there were not a fundamental, graceful simplicity embedded in the design and functionality of the Internet, it never would have become as pervasive as it has. While commonly used, easy-to-operate Internet applications, like Web browsers and email tools, keep the Internet's underlying behavior and technology effectively hidden from view, this doesn't necessarily mean that the operations happening behind the scenes are complex or difficult to grasp. As you'll soon appreciate, all of the various events that occur when you press that button to send off an email or submit a Web form can be described in plain language, largely because they consist of a series of simple, albeit quickly occurring, operations. That's because computers are remarkably simple, and even stupid, devices. They are ideally suited for performing certain jobs, like executing basic, highly repetitive tasks in a consistent and reliable manner and carrying out simple or complex calculations according to preset rules or algorithms with amazing speed and accuracy. But when it comes to any sort of reasoning or extensive, deductive thinking, or even to carrying out a complex sequence of instructions — the sorts of things we do routinely and innately — computers need everything defined down to the smallest detail, including how to recognize and handle unexpected events. The simpler their instructions, the better they operate. These instructions with respect to networking in general and the Internet in particular are an excellent example of why simplicity is so important to achieving success where computers are concerned.

Presenting the full story of the Internet — a review of its history, an explanation of its technology, and an analysis of its use, abuse, and impact — would be too much material to cover in a single book. For this reason, this book is the first in a series of three books; each book presents a distinctly different perspective on the Internet, as is explained here.

This book covers the essential history and technology of the Internet. It describes who built the Internet and why, how the Internet worked in the early days and how it works today, and what it meant for the Internet to be transformed into the private and commercial international network we use today after operating for twenty years as a scientific and military network owned and controlled by the U.S. government. It begins with a brief overview of the history of computers and of the evolution of their technology, use, and impact before the revolutionary introduction of networking. It continues by presenting the story of the very first computer network, the ARPANET, which was the direct result of a a government-sponsored technology race (not unlike the space race) following the launch of Sputnik. It then describes the Internet's early history and its coexistence with the ARPANET, the changes in the Internet's technology and management as it made its transition into today's Internet, and the details of how data makes its way through the Internet. This book also describes the protocols that constitute the essential technology of the Internet and it examines in detail several of the Internet's most popular services, including email, chat, gaming, and the virtual reality of cyberspace.

The second book in the series is entitled "The Information Revolution: The Not-For-Dummies Guide to the History, Technology, and Use of the World Wide Web." It focuses on the predominant information service on the Internet, which many people mistakenly equate with the Internet itself: the World Wide Web. Much of this confusion stems from the media, which routinely uses the two terms — the Web and the Internet — interchangeably. This misunderstanding is also a direct result of the immense popularity of Web browsers, which conveniently package several of the Internet's most popular services, including email, chat (or instant messaging), and FTP (a file transfer program), into a single Web-centric application. Browsers give the mistaken impression that the Web has largely subsumed or displaced the Internet. But, as you will discover, nothing could be farther from the truth. The Web is simply a service on the Internet. In terms of the Web's relationship with the Internet, it differs little from dozens of other Internet services. Its purpose is to provide an information space on the Internet that enables information objects — in the form of HTML files and files of photographs, illustrations, and other types of stored information — to be located, retrieved for viewing, and published for others to find and view.

"The Information Revolution" begins with an exploration of the information management services that were available on the Internet before and during the arrival of the Web in the early 1990s, which includes a short history of hypertext, the technology employed by the Web to interconnect its information objects regardless of their location (e.g., the clickable links in Web pages). This is followed by the story of the Web's creation, including the story of its creator, Tim Berners-Lee, and his commitment to ensuring that the Web remain free of the controlling influence of any single country, corporation, or organization. The mechanics of the Web's operation is then explained by focusing on the Web's three core elements: the HyperText Markup Language (HTML), the language of Web pages; the HyperText Transfer Protocol (HTTP), the protocol that defines how Web pages are retrieved; and the Uniform Resource Locator (URL), the standard that defines the addressing mechanism for Web objects as well as other Internet objects. Other topics include a description of what constitutes today's information Web, the influence of multimedia on the Web, e-commerce, relationship management, information filtering, and how the Web is managed. The book ends with an description of the efforts under way to overcome the known limitations of today's Web. The objective of these efforts is to enable the Web to evolve from an information space that is difficult to categorize and map into a knowledge space that is more easily known and easier to navigate. This knowledge space has been named, by the Web's creator and others, the Semantic Web.

The third book in the series is entitled "The Technology Revolution: The Not-For-Dummies Guide to the Impact, Perils, and Promise of the Internet." It explores subjects related to our use of the Internet. These subjects include how the growing integration of the Internet into our daily lives — at home, school, and work, for pleasure, education, and business — has changed the way we communicate with one another and how we conduct our lives. Also covered are the critical concerns stemming from our increased use of the Internet (e.g., the privacy and security of our personal information), an examination of the digital divide (i.e., who is on the Internet and who is not and what this means for all of us), and issues related to children's use of the Internet. This book also presents the history and technology of Java, a programming language that has evolved much as the Internet has evolved, and the story behind the dot com boom and bust of the 1990s, including my personal account of managing the software engineering and development department at a small startup dot com. Also examined in the book is how our use of the Internet is helping to redefine our concept of community by establishing the Internet as its own unique and powerful form of community. The book concludes by taking a long and wide perspective on the Internet and examining where the Internet fits with respect to the evolution of technology, the history of communication, the history of information storage and access, and the empowerment of the individual.

## Why Read This Book?

You have a child who seems to know everything about computers and the Internet, while you know little or nothing. Consequently, you've relinquished control of the computer to your child under the assumption that he or she knows how to behave on the Internet and uses all the appropriate precautions. Your company's reliance on the Internet is increasing all the time and you feel like the technology is often hindering your performance rather than helping it. You simply want to feel more comfortable with using your computer to access the Internet. You're interested in pursuing a career in computer technology, or perhaps in the Internet itself, and you want to understand, from the ground up, how it all works, or you want to learn how and why the technology was created in the first place. You have an interest in current events and in the factors influencing today's social structures, business trends, education, and economics, and you want to understand how much of an influence the Internet has become. This book will satisfy all of these needs and interests, and many others, by providing you with a complete, working knowledge of the Internet and how it is being used, along with an explanation of many of the fundamental operations of computers.

This book is not, however, a manual on how to use the Internet. It does not explain how to configure your computer to connect to the Internet, how to operate a Web browser or send email, how to download files or register a name for Internet chatting, how to configure virus detection software to protect your computer, or how to install and maintain content filtering software to protect your child. Unlike other Internet manuals or guides, this book will, however, answer your questions about what the Internet consists of, how it is managed, how data travels through it, why it was created, where it came from, and where it may be heading. From it, you will learn why and how computers store information and communicate in a digital language of zeroes and ones, what is occurring on your computer when a file is downloading, how email messages are sent and delivered, and why the technology of the Internet has been such a remarkable and unequivocal success.

For most people, the Internet is regarded as a black box; little or no effort is made to understand what is in the box or how it works. The availability and operation of the Internet is taken for granted; its technology and any consequences of using the technology when connecting to the Internet, sending and receiving email, browsing the Web, or communicating in a chat room are ignored or overlooked. This is as true for most casual Internet users as it is for the majority of people who use and rely on the Internet on a daily basis. It is also true for many of the people who make their living in the computer and Internet fields. One objective of this book (and of the other books in the series) is to help remedy this situation.

This book describes how the Internet evolved and how its layered, decentralized design led to the creation of remote computer access, email, instant messaging (or chatting), Internet gaming, the World Wide Web, and more. Central to the discussion of each component of the Internet is that component's individual history along with enough technical detail to provide a basic understanding of how the technology is implemented. After reading this book, you will understand exactly how the Internet and the Web differ. Moreover, you will recognize how they relate to each other, that is, how and where their technologies intersect and how each has influenced the other. You will also understand why a whole new type of communication system had to be engineered and built — one very different from the existing and very reliable telephone network — in order to accommodate the unique requirements of how computers communicate and the type of information computers typically exchange. Above all, you will appreciate what a remarkable and unique achievement the Internet is, how far its influence extends, and how the grassroots efforts of individuals were indispensable to making this achievement possible.

You don't have to understand how the Internet works to enjoy communicating via email or instant messaging, to browse the Web in search of the best price for a new or used car or to identify the bird that has taken up residence at your bird feeder, or even to create an Internet presence for your business or your family. But a little understanding will give you some confidence in pursuing these activities and many other activities just like them. Even better, a more thorough understanding will enable you unleash the full potential of the Internet, which many people have found to be an empowering experience. This knowledge can help you put the Internet to work for you, your business, and your family in ways you never would have imagined possible.

At the very least, you should understand something of how the Internet works and what it means for you to go online in order to make your experience surfing the Web and exchanging email a safe one. No matter how it feels, your presence on the Internet is never anonymous or passive. As you point-and-click your way across the Internet, information about who you are and what you are doing is being tracked and recorded all the time for various and sundry purposes. Your best protection against some sort of mishap or you, your children, or your business becoming a victim of some malicious act or criminal activity is a sound understanding of how the Internet works and what it means to venture into cyberspace.

The history of how and why computer networking was invented, the struggle to convince others early on of the significance and potential of networking, and the transformation and consolidation of those early networks into the Internet compose one of the most influential and critical technology stories of our time. When you know the Internet's history and then explore its impact and use, you will start to appreciate the larger social story that the Internet's creation, evolution, and popularity has to relate about us, about the lives we are leading today, and about what we may expect from this technology tomorrow. Moreover, when you understand the basic mechanics of how the Internet works, such as what is required for two computers to talk to each other, how data travels across the Internet and finds its way to its destination, and what is happening when two or more people are chatting online or playing a game from remote locations separated by a few miles or several thousand miles, you will suddenly feel empowered to put the Internet to work for you, safely and confidently.

# Computers: The Instrument of a Revolution

**1**

## The Internet's Foundation

Like the birth of Athena, goddess of wisdom, who sprang fully grown and clothed in her battle armor directly from the head of Zeus, it seems to many that the Internet just appeared one day out of the clear, blue sky or from some high-tech ether. When you read on and learn about all the surmounted obstacles, innovative inventions, and insightful engineering that made the Internet possible, you will appreciate that today's Internet represents one moment in both a long and fascinating history of scientific discovery and an associated, equally fascinating evolution of computer engineering and technology. Hopefully, before you finish this book, you will also appreciate how Athena, who held dominion over all things connected to the intellect and was considered the very embodiment of wisdom, has effectively found her way into our times in the form of the Internet.

In order to examine and understand the workings of the Internet, you first need to acquire a basic understanding of how computers operate. This is because computers represent the literal, and figurative, building blocks of the Internet. They also provide our access to the Internet. It is through computers, from wireless, handheld devices and laptops to supercomputers the size of refrigerators, that we interconnect with the Internet to send and receive email, browse the Web, place a bid for some impossible-to-find, automobile radiator cap, or meet up with friends in a chat room. It is also through computers, in the form of routers, gateways, switches, network hubs, and other specialized minicomputer hardware, and in an assortment of desktop computers, workstations, and servers, that the information we send and receive travels the globe from location to location, effectively forming the infrastructure that is the Internet.

As recently as the 1960s, however, computers were no more capable of communicating with one another, sharing information, or even transferring a simple file from one computer to another than typewriters or adding machines were able to share or exchange information. There was no Internet, nor was there any form of computer network. Few people even thought that such a widely divergent future for computers might exist, given the state of computer technology and the mindset regarding computers and their use that existed at the time. That's because the greatest obstacle to building a network and enabling computers to talk to one another and to share information was the computers themselves. Building the first network, therefore, began with transforming these computers, not simply in how they functioned and what they could do, but, more importantly, in how they were designed, engineered, and built.

For most of their history, computers consisted of adjunct devices, largely mechanical tools for performing calculations faster and with greater ease than we could perform those same calculations, and storing and manipulating large quantities of information faster and more efficiently than we could on paper. Over the years, their technology improved and they became capable of processing more data and of performing ever faster operations. What did not improve, however, was the limited way in which we interacted with computers. We worked with computers on their terms, adapting our behavior to the limitations of their engineering and technology. Moreover, they remained at a distance, both figuratively and literally, and performed their work in isolation from us and from each other. Once these limitations were recognized, addressed, and overcome, however, and the isolation of computers was broken and we could interact with them directly, all sorts of new and powerful applications for their use emerged as a result. As we will see, the pursuit for creating the world's first computer network began the process and established the environment for these changes to occur. Ultimately, these changes led directly, and sometimes indirectly, to the Internet and to its consequent revolution in information access, communication, and social interaction.

The transformation, or evolution, of computers and their technology continues unabated today. While the Internet has become the focal point of this evolution, a quick glance at our surroundings should confirm to what extent computer technology has pervaded our home and work environments, and has even become part of our recreational activities. Computers and their paraphernalia consume our desk space. Specialized computers monitor and adjust the performance of our cars and run the cable and satellite boxes that connect to our television sets. Smaller computer components control, and sometimes interconnect, many of the other electronic devices that have become commonplace fixtures in our busy lives, such as our telephones, beepers, exercise equipment, and household appliances. But nowhere is the influence of this technology more pronounced or its development more profound than in the creation and growth of the Internet.

No single event or invention is responsible for the Internet's existence or its popularity. What you will discover, however, in the sections below and in the chapters that follow, is that changes in the technology of computers combined and interacted with changes in how computers were perceived and with new, innovative thinking regarding the purposes that computers could fulfill to bring computers into our lives and, in turn, to change our lives forever. At the heart of all these changes is the Internet.

# A Short History of Computers

A computer is generally defined as a machine for automatically performing calculations. While this may sound like a definition better suited to a calculator, it nevertheless aptly expresses the primary function of all computers: however simple or complex, small or large, inexpensive or costly, every computer strives to complete some calculation or operation and report back either true or false, zero (0) or one (1). This is something of a simplification, but it emphasizes the point that computer intelligence is programmed intelligence; and every action we take on a computer evokes some predefined, corresponding action, or series of actions, within the computer along with a response. More specific definitions describe a computer as a programmable electronic device used in the storage, retrieval and processing of data. This sounds more like the devices that most of us use, because this definition takes an external perspective, as opposed to the internal view presented above. It characterizes the device by how we employ it, at the expense of describing what a computer consists of or how it works.

Ironically, an earlier definition for a computer describes not a mechanical creation of human beings, but a person. More specifically, a computer was once an expert at performing calculations, or an expert at operating calculating machines. During World War II, the very first mechanical computers that could be considered precursors of the computers we use today were just being invented. They were used primarily in the highly mathematical process of decrypting enemy ciphers and assisting in the difficult task of trying to break enemy codes. At the same time, greatly outnumbering these giant boxes of wires, wheels, and relays, countless human computers, mostly women, were using pen and paper along with their intellect to perform the same complex, mathematical tasks.

The history of mechanical computers dates back roughly another hundred years. It begins in 19th century England with mathematician and inventor Charles Babbage. Regarded by many as the *father of computing*, Babbage took an existing technology, the Jacquard mechanical loom, and derived from it an utterly new way to mechanize, and in effect program, analytical calculations. His goal was to create a device that was automated, general purpose in scope, able to execute repeated instructions in any specified order, and able to solve any algebraic equation.

Jacquard looms operate in the 21st century much as they did in the 18th century, using a chain of punched cards as a mechanical means to automate the weaving of complex patterns. While Babbage was never able to build his *Analytical Engine*, he described in detail how punched cards could be used to guide his machine through a prescribed series of calculations and how, by means of a second set of cards, variables could be introduced into these calculations.[1] The storage of instructions in one place and data in another, while simple in concept, is profound in its impact. Every computer program ever written builds on this notion. Even today's popular programming paradigm of object-oriented code re-use and modularity can at least in part be traced back to this simple, powerful concept.

Babbage's design also included a *store* to hold numbers and intermediate results as well as a *mill*, a separate area where arithmetic processing was performed. These separate components correlate to a modern computer's memory and central processing unit (CPU), respectively. Just like today's computers, his engine was meant to repeat operations, or loop, any number of times and to perform conditional operations by testing a condition and following a specific course of action based on the result.

Late in the 19th century, in time for the 1890 United States census and roughly twenty years after Babbage's death, Herman Hollerith created the first punched card tabulating machines. These early mechanical computers were in many ways the first realization of Babbage's Analytical Engine. By punching holes in cards, Hollerith was able to transfer the census data that had been collected for an individual, store it once, and re-use it any number of times. Each card served to identify an individual's gender, profession, and other characteristics. The tabulating machine would then be configured (or programmed) to check for one or more characteristics, the cards would be loaded, and as the cards were fed through the machine a counter would increment each time the programmed conditions were met. The counters were electrically operated, making the machine electro-mechanical in nature. A hole in the card would allow a spring loaded pin to go through the card when the programmed condition was met; and the pin would then make contact with a metal surface or metallic liquid below, thereby completing the circuit and causing the counter to increment.

While today Hollerith's tabulating machines may sound simple and strikingly low tech, they represent an enormous leap forward with respect to collecting and analyzing data and using that data to perform critical statistical analyses. For example, prior to Hollerith's invention, data from the census was barely processed within the ten year period decreed by Congress.[2] The use of his machines allowed the 1890 Census to save over two years and over five million dollars compared to the 1880 Census.[3] Moreover, a greater range of data was processed and with far fewer errors. This effort clearly marks the beginning of the age of data processing and of the kind of computer use we commonly see today.

Between the 1890 Census and World War II, Hollerith's tabulating machines and other mechanized computational devices found their way into businesses and government agencies in the United States and overseas. In 1911, Hollerith's own company, the Tabulating Machine Company, was merged with three other distinctly different companies to form the Computing-Tabulating-Recording Company, which would later be renamed IBM. Hollerith's machines were repeatedly refined during their extraordinary fifty year span of general use. But their core components and their design remained much the same. Their speed and capacity were increased; and the punch cards (also known widely as Hollerith cards) were enhanced to allow more data to be stored, which in turn led to more complex and sophisticated programming of the machines. In the end, however, their design and manufacture led to their demise.

Hollerith's computing machines were large, mechanical devices that required very specialized and regular maintenance, as well as a great deal of manual, human intervention. Furthermore, they were clumsy to operate and difficult to program. They busily and efficiently performed their assigned tasks, whether these tasks involved tabulating data from another census, or charting freight movements and revenues for the railroads, or compiling data for city health departments. In doing so, they effectively established and promoted a dependence on computing devices. Their technology lasted far longer than any of the technologies created to replace them. Unlike those later technologies, however, the technology of these early computers bears little resemblance to that found in the computers we use today.

An interesting example of these highly specialized, electro-mechanical computers could be found in the English countryside of World War II involved in what was, and remained for thirty or more years, the best kept secret of the war. Here, machines referred to as *bombes*, built by IBM's British counterpart, the British Tabulating Machine Company, assisted in the daily challenge of trying to find the encoding setting of the German's encrypting device called Enigma. The Germans considered the ciphers produced by their Enigma machine unbreakable. But by means of their considerable mathematical and analytical skills, their tireless perseverance, and with the help of these huge, elaborate, finicky computers, the British proved otherwise. Diana

Payne, one of the many women who joined the WRNS (Women's Royal Naval Service) and whose job it was to configure the bombes in accordance with a provided key, described them as follows:

> The bombes were bronze-coloured cabinets about eight feet tall and seven feet wide. The front housed rows of coloured circular drums, each about five inches in diameter and three inches deep. Inside each was a mass of wire brushes, every one of which had to be meticulously adjusted with tweezers to ensure that the electrical circuits did not short. The letters of the alphabet were painted round the outside of each drum. The back of the machine almost defies description — a mass of dangling plugs on rows of letters and numbers.[4]

## First Generation Computers

The origin of the computers we see and use today dates back to the 1930s and 1940s and the creation of the first generation of electronic, stored-program computers. (Most contemporary computers are generally considered fourth generation computers, as is explained below.) Electronic, at the time, meant vacuum tubes, which were used in place of the mechanical relays common in earlier computers. Prior to World War II, several different organizations worked independently to try to build faster, program-controlled calculating devices for scientific and engineering applications.[5] But the onset of war quickly altered the application and urgency of this new technology. The overall objectives, however, remained the same. In the end, the increased computational power and speed of these new computers proved critical to the creation of ballistics' tables, the design of weapons, general logistics, and the challenges of code breaking.

A computer called the ENIAC (Electronic Numerical Integrator and Computer) was built for the Army in 1943 and clearly illustrated the performance advantages of electronic components over their mechanical counterparts. It was considered a hybrid computer since it contained the newer, electronic technology of vacuum tubes while retaining some of the electro-mechanical relays of the older machines. The ENIAC was huge, incredibly complex, and consumed a vast amount of power. It filled a large room and consisted of forty panels containing over 1500 electro-mechanical relays and 17,000 vacuum tubes.[6] The ENIAC could perform calculations hundreds of times faster than the purely electro-mechanical computers of the time. It did nothing, however, to reduce the time and labor needed to program a problem into the computer. Nonetheless, the U.S. patent office recognized the ENIAC as the first computer.

The breakthrough in computing architecture came a few short years later at the University of Pennsylvania's Moore School of Electrical Engineering. A draft report describing progress on the EDVAC (Electronic Discrete Variable Arithmetic Computer), written in 1945 by mathematician John von Neumann and others at the Moore School, detailed the logical design of a stored-program computer. They divided the computer system into five units:[7]

- A *central arithmetic unit*, for performing basic and higher level math functions.

- A *central control unit*, for handling the sequencing of operations.

- A *memory unit*, for storing numerical data and numerically coded instructions.

- An *input unit*, for transferring information from some external recording device, a keyboard or computer tape, for instance, to the arithmetic, control, and memory units.

- An *output unit*, for transferring information from the data processing and memory units to some external recording device, such as a display or computer tape.

Like Babbage a hundred years before, von Neumann found the best approach to analytical design in discrete components that handled the core computer functions separately. His division of labor became the guiding principles for all subsequent computer hardware engineering. Moreover, it allowed programming logic and data to be stored within the computer, eliminating the slow and tedious process of loading programs from punched paper tape or manually through the keyboard.

Coincidentally, 1945 also marks the first recorded computer bug, literally. The heat and glowing light produced by the vacuum tubes in this generation of computers were perfectly suited to attracting moths. This first bug, therefore, manifested itself as a moth stuck between the relays on the Harvard Mark II computer. It was discovered there by Dr. Grace Hopper of the U.S. Navy. While computer bugs are common today and mostly manifest themselves as unwanted or unexpected behavior resulting from poor software engineering and inadequate coding, these computers provided the first, and very different, definition of computer bugs and the time consuming, often tedious process of debugging.

Considered the first business computer made in the U.S., the UNIVAC I was built in 1951 by Remington Rand with funding provided by the designers of the ENIAC, John Mauchly and J. Presper Eckert. The UNIVAC was used to predict the outcome of the 1952 United States presidential election and appeared on television alongside well known news reporter Walter Cronkite. The UNIVAC's appearance on the news, incidentally, made it the first computer most people had ever seen. Adlai Stevenson was considered by many the likely winner of the election based on the results collected through opinion polls. But hours before the polls closed on the west coast, the UNIVAC projected 100-to-1 odds that Dwight D. Eisenhower would win. Ironically, the results produced by the UNIVAC were not reported by Cronkite until hours later because they were not believed to be accurate. Just as the UNIVAC had forecast, however, Eisenhower won by a landslide. Consequently, computers and their potential applications immediately took on greatly increased significance.

## Second Generation Computers

In the second generation of computers, which appeared roughly ten years later, the concept of *computer systems* transformed the way computers were engineered, manufactured, marketed, and sold, as companies started to allow for different configurations of memory size, number of processors, input/output devices, and other components. Another evolutionary component key to the technology of this generation of computers was the transistor. The transistor, invented at AT&T Bell Laboratories in 1947, replaced the vacuum tube of the earlier generation of computers and was instrumental in creating these faster, smaller, more reliable, and more powerful computers.

Vacuum tubes were the first electronic switches. They could perform two basic tasks: amplify sound, which was essential to the operation of radios and television; and rapidly turn on and off, roughly 10,000 times a second, which was essential to the operation of digital computers in which everything resolved to on or off, true or false, zero or one. Vacuum tubes were large, produced heat, burned out frequently, consumed a lot of power, and required constant maintenance. The transistor, on the other hand, was much smaller, was far faster in switching its state (i.e., turning on or off) even in its earliest models, consumed less energy, and was extremely reliable. Instead of using a glass bulb with a filament and a vacuum, the transistor performed the same two functions as a vacuum tube by moving electronic charges inside a solid block of semiconductor material, such as silicon or germanium, along controlled pathways. (Semiconductor material belongs to a class of solids that can conduct, or carry, electrical current while also acting as an insulator, ensuring that no, or minimal, heat is given off.) Unlike a vacuum tube, the transistor produced no heat, required no warm up time, and included no component that would burn out over time.[8]

Also key to these new computers was how they stored information. Their development charts the migration from cathode-ray-tube and delay-line memories to magnetic ferrite cores and magnetic drums. Just like the migration from electro-mechanical relays to vacuum tubes to transistors, this evolution in the technology of how to store information in the form of binary digits (zeros and ones) for later recall charted a path towards smaller, simpler, less expensive, more reliable components that performed the same basic operation. These second generation computers were also designed so that they could house more discrete hardware components to better divide up the computational labor. Additionally, they were engineered to function with new, high-level programming languages and associated software to simplify and enrich programming.

The combination of these new features also allowed for something called *batch processing*, a far more economical and efficient means of getting access to a computer in order to run a program. Programs could be stored in advance on magnetic tape; a computer operator would load the programs when computer time was scheduled or available and run out the results to magnetic tape. Computer time was costly and access was limited. Batch processing was a simple and effective way to maximize usage. It was also key to the debate surrounding the computers connected to the first computer network, called the ARPANET, which will be explained in the next chapter.

## Third Generation Computers

The third generation of computers appeared in 1965. Perhaps the most significant of these computers was IBM's System/360 series. The advances in these computers consisted largely of the following:

- Smaller, less expensive integrated circuits replaced discrete transistor circuits.

- Semiconductor memories (also made possible by the new integrated circuit technology) replaced magnetic drums and magnetic ferrite cores.

- The creation of microprogramming, also called firmware, provided a bridge between the computer's hardware and programming and thereby simplified circuit design.

- The creation of parallel processing, the execution of more than one instruction at the same time, increased computer efficiency.

- The creation of operating systems, master programs that controlled the system, scheduled jobs, allocated resources, and monitored operations, provided far greater control over the computer's operation.

The first integrated circuit was produced by Texas Instruments in 1958. It was a crude proof-of-concept containing a sliver of germanium with five components connected by wires. Also known as a semiconductor chip, an integrated circuit interconnected multiple transistors along with resistors (i.e., components designed to limit or regulate the flow of electrical current) on a single piece of semiconductor material, a chip. It solved, with striking simplicity, the fundamental engineering dilemma associated with using transistors that was generally referred to as the *tyranny of numbers*.

In order to build faster, more powerful computers you needed more circuits, which in turn meant more transistors. There was no limitation to the number of transistors you could wire together. There was, however, a limitation with respect to how fast electrical signals could travel that was impossible to ignore and impossible to overcome; and electrical signals were used to switch the state of each transistor between on and off. This limitation was the speed of light. Therefore, larger circuits with more transistors resulted in slower computational speeds. To increase computational speed you had to decrease the size of the circuits, which meant decreased computational capacity.

The integrated circuit solved this dilemma by reducing the number of parts in a complete circuit to one. Rather than wiring individual transistors together to create a particular circuit design, that same design would be implemented on a single piece of semiconductor material, such a silicon wafer, containing all the necessary components and interconnections in the smallest possible space. This revolutionary invention was hailed by the National Academy of Sciences as the originator of the *Second Industrial Revolution*. It signaled a new era in manufacturing that affected everything from clocks and toys to telephones and computers. But, more than that, it meant that computers could now take on any size or type of computational labor.[9] The co-founder of Intel, George Moore, observed that since the invention of the integrated circuit the number of elements held in each square inch of a chip was double that of the previous year. This comment quickly translated into Moore's Law: "no sooner has a computer been produced than it is out of date."[10]

This third generation of computers also included a departure in computer architecture development with the advent of minicomputers: smaller, less expensive devices more limited, and more focused, in their capabilities. Such computers were perfect for more narrowly defined operations and, not incidentally, were key to the creation and success of the ARPANET, as will be explained in the next chapter.

## Fourth Generation Computers

Fourth generation computers, the ones most of us use today on a daily basis, are distinguished by microprocessors. The first commercially available microprocessor was the Intel 4004, which was introduced in 1971. The microprocessor was made possible by advances in manufacturing that facilitated the production of integrated circuits with a density of hundreds to thousands of transistors or logic gates per chip. (Earlier chips were limited to no more than one hundred transistors or gates per chip.) These large-scale integration (LSI) chips evolved into very large-scale integration (VLSI) chips that were capable of fitting several hundred thousand transistors per chip. Since the production of integrated circuits is mostly automated and the cost per chip is relatively small, the creation of these VLSI chips revolutionized the computer industry, starting with the creation and mass production of affordable personal computers. Integrated circuits are also responsible for the more recent and more adaptable computerized technology we see and use everyday, but which remains more or less hidden from view. This is the technology, for instance, that is responsible for programmable cell phones, toy cats that meow and demand to be fed or petted, and smarter traffic lights that can adjust the timing of their signals based on changing traffic conditions.

What made the microprocessor a revolutionary advance in the development of computer technology wasn't simply the density of the transistors it could house. Rather, the microprocessor marked a critical change in direction with respect to the general architecture of computers. Instead of manufacturers producing ever more complicated integrated circuits designed to meet highly specific needs, the microprocessor allowed them to create general-purpose chips that could be coupled with programmed intelligence to meet any number of needs. This change can be seen in the advertising Intel used to market its first microprocessor, which it called a "computer on a chip."[11]

## The Human Element

Scientific discovery and technology advance at their own pace. These advances occur separate from us and separate from how we interact with (or choose not to interact with) their effects, influence, and the inevitable products that get created in their wake. More often than not, these advances occur in cycles. Some cycles are large and represent significant, if not revolutionary, change, such as the invention of the transistor or the integrated circuit. Other cycles are small and more incremental in nature, such as the ever increasing size of computer memory and general, continual increases in computational speed. What we can never know, despite the best intentions or the most reassuring guarantees, is where our technological inventions will take us. The Internet, as you will see, is a case in point.

The first half of the 20th century amply demonstrated the value of computers. Government, business, and the scientific community all discovered how to employ these new, technological wonders to perform existing work faster and cheaper, and to accomplish new things. World War II accelerated the advancement of computer technology as the United States and other governments funded computer research projects and employed computers for everything from atom bomb research to weather predictions. When the U.S.S.R launched the first satellite, Sputnik, into space in 1957, a new technology-based race was started. The participants and the competitors have changed over the years, but the race itself has never ended. This race marked a turning point in the role computers would play in our lives, transforming them from remote, calculating machines and islands unto themselves into familiar, personal productivity tools and electronic filing cabinets interconnected across an endless sea of like-minded computers. However unintentional the effect, the human element was introduced into the world of computers. The Internet was an early and lasting consequence.

Consider for a moment how computers are routinely used today. For one thing, we rely on computers to communicate; and, for some people, this represents their sole purpose for having a computer. Common ways people communicate through their computers include sending email, chatting online, participating in remote business meetings, and even making phone calls, activities made possible by the Internet, incidentally. All of these commonplace and fundamental current computer activities are predicated on the assumption that computers can easily exchange information, and that some network exists to carry and control all this communication traffic. These activities are also based on the assumption that computers were designed to function and interact much as we behave and interact. In other words, we assume that the computers we use today have been designed with our specific needs in mind and adapted in form and function to suit how we live and work.

But, as recently as the 1960s, such activities were confined to the realms of wishful thinking and science fiction. Few people considered the possibility that computers might some day be used for the purpose of communication, or that it would be common for individuals to own personal computers. Standing in the way of any practical implementation of such ideas were both technological obstacles (great and small) and the mindset of the time, which regarded computers and human-computer interaction in ways dictated by how the computers of the time were designed, built, and accessed.

Computers were scarce, very large and expensive, and awkward to use. Computer networks existed only on paper. Computers were still primarily computational devices with which very few people interacted and which worked in isolation. Even a simple task like sharing or transferring a piece of software from one location to another took considerable time, energy, and planning since you needed first to place the program onto magnetic tape or punch cards and then carry it from one location to the other. Shoe boxes, incidentally, were the preferred means of transporting and storing punch cards on college campuses, even well into the 1970s and early 1980s. It wasn't just pocket protectors (sleeves of plastic that slid into shirt pockets to prevent pen ink from leaking onto a shirt) that helped computer enthusiasts of the time stand out from the crowd.

In the pre-network days of the 1960s, the typical programming model was quite laborious and relied on the batch processing approach for programs to be executed by the computer in a timely and economical manner. The process was both frustrating and inefficient, especially when compared with how we do programming today, our individual and largely unrestricted computer access, and the variety of tools we can use to test and debug our programs. When it was introduced, however, batch processing was a critical advance towards both granting wider access to computers and maximizing computer use. The following sequential steps illustrate how the basic batch processing model operated:

1. A programmer wrote a program out on paper.

2. A programmer or keypunch operator transferred the program to punch cards thereby creating a code the computer could read.

3. Cards were transported to the computer center and fed into the card reader, transferring the program onto magnetic tape.

4. An operator loaded the tape and ran the program when computer time was made available.

5. A printout of the results was returned to the programmer.

6. If errors were found (more often the case than not), the whole process would need to be repeated, frequently starting from step one.

Typically, this cycle was repeated many times before a program would run through completely. A single logic error, typographical error, or mis-punched hole would cause the program to terminate prematurely.

While batch processing was an expedient way to adapt human behavior for more efficient and cost effective computer usage, some computer scientists of the time were set on reversing this equation. They focused on adapting computers and how they were accessed along the lines of how humans posed and went about solving problems; and in so doing they expected to achieve the same efficiencies as batch processing while greatly reducing existing programming time and overhead. They called their model *time-sharing*.

A major proponent of time-sharing was J. C. R. Licklider. Licklider, along with his colleague Welden Clark, presented a paper in 1962 in which they explained how several people could use a computer at the same time. While one person was working on a logic problem or doing some other preparatory work, the computer could be performing calculations for someone else, thereby reducing the idle time of the computer. Key to time-sharing was that there would be multiple terminals connected to a single computer. Each terminal would appear to have immediate and complete control of the computer. But, in actuality, the computer would be swiftly alternating focus from one person's request to another's. This radical approach of allowing direct and immediate access to a computer made possible new and powerful uses for the computer that previously would have been inconceivable. Computer-based learning was one simple example. A network of interconnected computers that people could access remotely represents a more extreme, far-reaching example. Not coincidentally, both Licklider and time-sharing played an important role in the development of the ARPANET, as will be presented in the next chapter.

In terms of our interaction with computers, this reversal of the equation between computers and humans, namely adapting the design and functionality of computers in accordance with our behavior and needs, is what began the process of bringing computers into our everyday lives. This particular effect was still a couple of decades and many amazing technological creations away, but it began at that very moment, with the introduction of time-sharing and the reversal of this human-computer equation. It began when we stopped looking at computers as something adjunct to our lives, as machines in the periphery that performed their assigned tasks and then handed us back the information we needed to continue with our work or support our conclusions. It began when computers started to be viewed as aids and extensions to how we went about learning, how we accessed and used information, and, not least of all, how we communicated. In other words, the human element was introduced. As a consequence, both computers and our lives were forever changed in the process.

# The Missing Link, Literally

Where would computers be today without the existence of networks? It's one thing to engineer computers based on how we process and interact with information. But most of what we do and how we live our lives revolves around our interaction with others. In order for computers to become integrated into our lives they had to enhance and extend this fundamental part of our existence. Networking made this possible.

All things considered, computer technology evolved quickly over the course of the 20th century. Even so, computers were, and still are, machines that process information, the information we feed them. Each successive generation has operated some order of magnitude faster with respect to this raw processing power. Meanwhile, computers now come in all sorts of shapes, sizes, colors, and configurations. The diversity is striking given how fundamentally similar they all are in terms of receiving, processing, and outputting information. The differences, in most cases, are only skin deep. This is especially true when it comes to their networking components. Computer networking, much like human communication, is dependent on a common language and the constructs that accompany language that tell computers how to greet each other, interpret what is being communicated, and say good-bye. But up until about the time of the first lunar landing in 1969, when the ARPANET established the first working network by interconnecting its first four computers, computer networking was the missing link.

Today it's hard to imagine computers as remote objects, isolated, existing only in some climate-controlled secure room or computer center and accessible only from that location. We take networking for granted, whether it consists of connecting two computers on opposite sides of an office cubicle into a company's local, internal network (its intranet), or interconnecting a printer and a couple of computers at home, or enabling us over a telephone line to reach the Internet and interact with a computer on the far side of the country, continent, or globe. Computer networking literally changed everything: how we interact with computers, the functions that computers can perform, and, most importantly, how we interact with each other. What started out as the networking of computers quickly became the networking of information and people. In considering again how the definition of a computer has changed over time, one has to wonder when its definition will include mention of its role in networking.

# The ARPANET: A Network is Born

**2**

## Discovering the Road to a Network

The ARPANET was the first packet-switched network of computers. Its purpose was largely experimental; and during its twenty years of operation, from 1969 through 1989, it provided both a testbed and a proving ground for much of the networking technology and many of the most significant computer-based services we commonly use today. For example, email and instant messaging (or chatting) were created and developed on the ARPANET.

The Internet so many of us access today, the Internet we rely on in our personal lives and at work, evolved from the ARPANET. More specifically, today's Internet represents a sort of inheritance left to us by the many creative people who built the ARPANET and who recognized both the vast potential of the ARPANET and, more generally, the inherent power of interconnecting computers by means of a network. This chapter presents the history of the ARPANET and describes both the obstacles that had to be overcome to create it and the technology and engineering that made it a success.

## Sputnik

The story of the ARPANET begins with the Soviet Union's launch of Sputnik. On October 4th, 1957, this small, but powerfully symbolic satellite, the size of a basketball and weighing just 183 pounds, became the first man-made, artificial object to orbit the Earth. Every 98 minutes, Sputnik completed its elliptical orbit, much to the consternation of the United States government and military. Although the U.S. was already a couple of years into its own satellite development program with the Naval Research Laboratory's Vanguard project, the successful launching of Sputnik took the U.S. by surprise. Before a month passed and the Soviets launched their second satellite, Sputnik II (which, incidentally, included a passenger, a dog named Laika), the U.S. Department of Defense authorized the start of another satellite program called the Explorer project. A little more than three months later, on January 31st, 1958, Explorer I joined Sputnik I and II in orbit around the Earth.

From today's vantage point, Sputnik I seems a relatively innocuous object. Even its name, translated as "fellow traveler," conveys its genuinely peaceful, scientific purpose. At the time there were many official U.S. statements that welcomed the peaceful achievement of the Soviet scientists and expressed the hope that outer space would never be used for military purposes. Nevertheless, Sputnik I did more than herald the start of the space age and the consequent space race between the U.S. and U.S.S.R. Given the existing paranoia about the spread of Communism and the possibility of nuclear attack, its direct effect on the U.S. military was profound. Indirectly, however, it led to the creation of the ARPANET, the very first network of computers, and its successor, today's Internet.

Note that as a testament to the Internet (and, not incidentally, to the social and political improvements made in the last forty years or so), it is now possible to visit a U.S. government Web site run by NASA and not only read U.S. *and* Soviet accounts of the launching of Sputnik and the technology behind the achievement, but also hear a recording of the sound produced by Sputnik's telemetry as it signaled its solitary journey above the planet. The networking of computers has not only made such access to information possible, it's also responsible for what this access means: the breaking down of the barriers that separate people, communities, and nations.

## ARPA is Formed

Less than a year after Sputnik I was launched, the U.S. Congress passed the Space Act, which created the National Aeronautics and Space Administration (NASA). Just prior to NASA's creation, however, the first direct response by the U.S. to Sputnik's launch occurred: the establishment in 1958 of the Advanced Research Projects Agency (ARPA). ARPA's mission, then as now, is stated as follows:

> to assure that the U.S. maintains a lead in applying state-of-the-art technology for military capabilities and to prevent technological surprise from her adversaries.[1]

This was, and remains, a deliberately broad mandate. From the start, ARPA's objectives and ARPA's funding were supplied by the military, a fact that was formally emphasized several years after the organization's creation when ARPA was officially renamed the Defense Advanced Research Projects Agency (DARPA). But what was good for the military was not necessarily limited to military applications. This is why the history of the ARPANET dates back to the formation of ARPA. It was at ARPA in the 1960s and 1970s that scientists drove forward many of the major technological advances in microelectronics, computing, and, not least of all, network communications. Accompanying these technological advances, ARPA scientists and engineers advocated and subsequently implemented a fundamental change in how computers would be used and for what purposes. This change began with the arrival of J. C. R. Licklider.

Licklider was one of the main proponents behind a new and innovative approach to computer engineering called time-sharing. In 1960, Licklider had published his seminal paper, "Man Computer Symbiosis," which presented a vision of interactive computer use that closely resembles what we enjoy today, with individuals having direct and immediate access to a computer for solving problems and performing other supportive tasks. Amazingly, Licklider's description that computers should be designed and used in accordance with how we worked, posed problems, and interacted, bore no resemblance to the computer environment of the 1950s or early 1960s, in which computers were isolated, accessed through indirect means, and exceedingly limited in both the nature and scope of the tasks they could perform. Achieving what Licklider described in 1960 would take two full decades of advances in computer hardware, software, and engineering. But soon after Licklider joined ARPA in 1962, as the head of a small department called the Information Processing Techniques Office (IPTO), he began laying the groundwork for this fundamental change in computer access and human-computer interaction to occur.

Licklider's initial assignment in the IPTO's Command and Control division involved applying a computer's analytical power to ever-changing battlefield situations. This work focused on feeding data that represented the conditions of a particular conflict into a computer; and then using the computer's analysis of the conditions to assist commanders in making their decisions.[2] Licklider quickly realized that the existing computer-interaction model, called batch processing, was grossly inadequate for the performance of such work. Constraints on time, combined with the highly dynamic battlefield environment, did not allow for the delays and discontinuity inherent in batch processing, which required handing off data to a remote computer operator, who then had to feed the data into the computer, run the program, and wait for the response, and, finally, communicating the results back to the command center or battlefield.

For Licklider, the only workable solution he could envision was predicated on providing multiple users with direct and immediate access to a computer, along the lines of the time-sharing computer-interaction model.[3] Time-sharing, however, was just theory. Plus, the notion of a single computer being accessed and shared by multiple users at the same time represented only part of the problem that needed to be resolved; and it was the the simpler part. The more difficult part was how to provide remote access, something crucial in leveraging a computer's computational power directly from a battlefield or some other, distant location. Solving these two problems evolved into the basic requirements behind the creation of the first computer network, an interconnection of ARPA's host computers distributed across the U.S., mostly at university locations, and named, appropriately enough, the ARPANET.

The effort to create the ARPANET would take seven years and, as described below, require the invention, engineering, building, and testing of both new computer hardware and entirely new methodologies for capturing, interpreting, and communicating information. But, shortly before the 1960s ended, the very first network operation took place on the ARPANET. This operation, while simple and brief, consisted of remotely accessing one computer from another computer's terminal. But it demonstrated precisely what Licklider and the military needed to achieve in order to analyze a battlefield situation remotely and as the battle was progressing. Plus, on a larger scale, this simple operation was key to any successful implementation of time-sharing.

During the two years that Licklider remained at the IPTO, he assembled a team of computer scientists to study the feasibility of the time-sharing computer model. He did so by contacting the leading research institutions in the U.S. and funding specific projects at those institutions through research contracts. In all, he had roughly a dozen universities working on ARPA contracts, including Stanford and UCLA where the first two ARPANET sites would be located. Licklider nicknamed this diverse and distributed group, composed of IPTO personnel and university scientists, the Intergalactic Computer Network. In a 1963 memo to the group, Licklider described the goals that would drive the work of the IPTO for many years to come. This work included computer system standardization and the interconnection of computers across an integrated network, both of which would help individuals share information and build on each other's work. He even described a type of software that would be machine, or computer platform, independent. Such software would be compiled once, stored on the network, and be ready for use by any computer on the network. (It would take nearly thirty years before the creation of Java at Sun Microsystems would realize this particular goal.)

Licklider's ideas and leadership advanced concepts and research that were key to both the formation of computer networks and the transformation of computers into the sort of multiuser, multitasking devices we see today. In pursuing this type of work with his Intergalactic Computer Network group, he effectively changed the department's focus from resolving strictly military problems to promoting general research into advanced computer techniques. Today, over forty years later, the work that Licklider and the IPTO began at ARPA continues unabated, as the IPTO's current mission statement manifests:

> [The] IPTO will create a new generation of computational and information systems that possess capabilities far beyond those of current systems. These cognitive systems — systems that know what they're doing:
>
> - will be able to reason, using substantial amounts of appropriately represented knowledge;
> - will learn from their experiences and improve their performance over time;
> - will be capable of explaining themselves and taking naturally expressed direction from humans;
> - will be aware of themselves and able to reflect on their own behavior;
> - will be able to respond robustly to surprises, in a very general way.

Today's mission statement, with its focus on artificial intelligence, is a far cry from the issues faced in the 1960s, when computers were largely islands unto themselves and we had to learn to speak their language and adapt our behavior to their limited functionality in order to put them to work. Even though the specific objectives have changed, the IPTO's overall focus of identifying and pursuing tomorrow's technology remains very much the same.

When Robert Taylor took over as Director of the IPTO in 1966, the demand for computer resources had been steadily increasing from year to year, but ARPA was no closer to establishing any kind of functional computer network. Taylor was immediately confronted with the following conditions:

- Three separate computer terminals on his desk, each connected to a different computer. (At the time, no single terminal existed that could be connected to, and communicate with, all three computers.)

- Requests from every location to purchase the latest computer technology.

Taylor's solution was as elegant as it was simple: he mandated that every location, including those at the research institutions, would have its own computer, but the locations would share their resources by ARPA building a computer network across the country to interconnect all the different sites and their resources.[4]

Taylor's bold mandate communicated the need for a network in clear and concrete terms. His belief, not unlike Licklider's, was that resource sharing represented the future of computing. The creation of a network was the first step in reaching out towards realizing that future. Like President Kennedy's speech that set the goal for landing men on the moon by the decade's end, Taylor's mandate stated the goal for computer networking in a way that left no uncertainty about ARPA reaching that goal. In 1966, Taylor identified a destination on the road that Licklider had started ARPA down in 1962; they would reach that destination at the end of 1969 with the ARPANET's first transmission of data across its network. Before the network could be built, however, they first needed to define what it would consist of and how it would work. A lot of questions needed to be asked and answered before a wire could be laid or any devices could be built to form the network's infrastructure.

## The First Engineering Challenges

Two substantial obstacles stood in the way of creating any kind of network. The first was a matter of comprehension: how do you get two dissimilar computers to understand each other; and how do you get information created on one type of computer to be correctly interpreted on another? The second was a matter of transportation: how do you transfer data from one computer to another and, in the process, provide some degree of efficiency and reliability?

Unlike today, computer standards were the exception rather than the rule in the 1960s. This meant that each computer manufacturer was creating, and building its products to, its own, often unique specifications. Since there was no network to interconnect computers, computer manufacturers had no reason to give any thought to sharing information between different types of computers. This meant that each computer spoke its own language and no other. Imagine for a moment a United Nations in which each country's representative speaks only his or her own language. In order for any two representatives to communicate directly, at least one would have to learn the other's language to handle the necessary translation of information. Multiply this scenario by the number of countries and you discover a hopelessly complex situation. This approach would never fulfill the expectations of an environment designed to facilitate communication and the sharing of information between the representatives of several hundred countries. Apply the same conditions to the computers of the time. Even a small network of only a dozen or so computers amply illustrated the magnitude of this issue to the engineers at ARPA. Moreover, the situation would only grow more complex over time as additional computers joined the network.

As if the issue of getting two computers to understand each other wasn't enough, ARPA's engineers didn't fare any better when it came time to examine how they might transport data between computers. They needed some sort of network infrastructure to provide the physical interconnection between the computers and to handle this transfer of data. The only existing large-scale communication network of the time was the telephone network, which was based on a networking model called circuit switching. Circuit switching was designed to meet the needs of human communication by establishing on demand a dedicated connection between any two telephone locations. Unfortunately, a circuit-switched network that was suitable for human communication was considered inadequate to accommodate the needs of a network of computers. As we shall see in detail below, factors relating to the cost structure of a circuit-switched network, as well as the very different sort of demands made by computers on a network's efficiency and reliability, meant that the telephone network could not fulfill the infrastructure needs of a computer network. This was the situation encountered by Taylor and his ARPA engineers when they started to consider how any two of their computers might be adapted to communicate with each other.

The first suggestion on how to network the computers was offered by Larry Roberts, a computer scientist from MIT's Lincoln Labs whom Taylor had hired to direct the networking project. His idea was to attach a modem to each computer and use the existing telephone network to establish a link and transfer the data. The source, or originating, computer would make a telephone call to the destination computer in order to establish a direct, network connection from one to the other. The modems would operate just as they do today, translating the digital representation of data stored in the computer into and out of an analogous audio format so that it could be carried by the telephone network. Roberts had used just such an approach in networking experiments he had conducted while at MIT, which meant that the engineering was proven. This approach also had the advantage of using existing technology. It could, therefore, be implemented in a relatively short time frame. After some discussion, however, Roberts' suggestion was rejected. The telephone network was considered too costly to use due to the long distance telephone charges. More importantly, concerns were voiced about the likelihood of lost data due to telephone line transmission errors or failures. Plus, this approach failed to address the difficult task of programming each computer to communicate with the other computers on the network.

Shortly afterwards, an engineer named Wesley Clark proposed a radically different approach to Roberts in the back of a taxi on a trip to the airport. Clark suggested that, if small, dedicated computers were used to handle, or route, the traffic of data on the network, each of the networked host computers would only need to learn a single language in order to join the network, the language needed to communicate with the dedicated, routing computers. These smaller, routing computers, or minicomputers, would be called interface message processors (IMPs). Here was born the notion of a *subnet*, a physically independent network segment that simplified the interconnection of computers on its network and allowed for the interconnection of it with other networks. Subnets became an essential and integral part of the ARPANET and of nearly every network that has followed, including the Internet.

As for how data would be transmitted across the network, using the existing telephone network was, as we have seen, not considered feasible. The communication needs of computers were in many respects the opposite of those of humans and could, therefore, not be accommodated by circuit switching. Consequently, it would be necessary to design and build a new and very different type of communication infrastructure. Fortunately, Roberts found his solution to this problem while attending a computing symposium in Gatlinburg, Tennessee. It was there that he learned of a technology called packet switching that transferred data in a way analogous to the way in which computers handled data.[5]

The history and significance of packet switching will be described at length in chapter 4. What's important to understand now is that packet switching offered a way to distribute the cost of transferring data while providing a reliable and efficient means of dividing the data into smaller units, or messages, in order to maximize the use of the telephone lines. Packet switching also provided an important capability sought by the military: a survivable communications network. As we shall see later, packet switching allowed the network to be both distributed and redundant, which meant that it could continue to operate even if some or most of it was destroyed. But packet switching was still only theory and unproven engineering; no one had yet built a packet-switched network. The ARPANET would be the first.

The following list summarizes how packet switching meets the needs of a communication network for computers by contrasting the features of packet switching to those of the circuit-switched system used for placing telephone calls.

- Packet switching communication is asynchronous: a computer knows when a packet of data has been sent but not when it has been received, which is sufficient for most types of data transfers. With circuit switching the connection is synchronous: the line is reserved for your use and communication back and forth is immediate.

- Packet switching allows for the efficient sharing of network resources: a computer can remain on the network continuously, but it only consumes resources while sending and receiving packets of data (typically a small percentage of the time it is on the network), and it transmits messages for any number of people. With circuit switching the connection is held open continuously and reserved entirely for your use, whether you are speaking or not.

■ Packet switching guarantees the delivery of data, trying repeatedly and through any number of pathways to complete a transmission. With circuit switching there is no guarantee that someone is there to pick up the telephone at the receiving end.

■ Packet switching slows as traffic increases, but does not compromise transmission quality or reliability. With circuit switching traffic congestion may cause the quality of the connection to be degraded or the connection to be interrupted, and a point may be reached at which resources are no longer available.

The road to the first computer network had been discovered, on paper at least. Host computers would only appear to be communicating directly with each other. In actuality, data from the hosts would be turned into packets at an IMP minicomputer, sent across the packet-switched IMP subnet, and finally reassembled at another IMP before being transferred to the destination host computer. The hosts would remain purposefully oblivious to the networking operations of the IMPs, minimizing the effort needed at each host computer to attach them to the network. Meanwhile, the IMP subnet would conveniently perform all the necessary networking functions and resolve the issue of comprehension, transferring the data efficiently and reliably from location to location and absorbing the language differences between the various types and models of host computers.

By 1967, just nine short years after Sputnik I started transmitting its signal, the study of computer science had essentially been created (thanks in large part to the funding supplied by ARPA to their university-based projects), the building blocks of the first computer network had been engineered and were awaiting development and testing, and both a survivable communications network for the military and a computer network to share expensive resources and reduce operating costs were in sight. All the ARPA computer scientists had to do now was figure out how to build the network and then see what would happen when they turned it on.

# Engineering and Building the First Network

Roberts possessed a clear vision of what he hoped the ARPANET would one day accomplish. It would, he believed, bring researchers closer together by facilitating communication and cooperation. He envisioned computers being used as community resources and enabling something he called *cooperative programming*, which would allow geographically separated people to work collectively, sharing resources and solving problems due to the interconnections fostered by the network.[6] Roberts, like Licklider, believed that time-sharing was the future of computing and that an efficient, reliable network would go a long way towards making time-sharing a reality.[7] However, as is true of much of the history of the ARPANET, the Internet, and the Web, what occurred over time ended up being quite different from what was sought and anticipated. Neither a cooperative programming environment nor time-sharing were ever realized with the ARPANET. Instead, the network took on a life of its own, developing an infrastructure, technology, and grassroots spirit that later evolved into the Internet and led to an unanticipated revolution in both communication and information access.

By June of 1968 Roberts had a formal plan for the development of the ARPANET and the funding to get started. Of the three main components that would compose the ARPANET - the IPTO host computers, the computer packet switching nodes (IMPs) that would route the data packets from one part of the network to another, and the leased 56 Kilobits-per-second (Kbps) telephone lines to connect the switching nodes — only the switching nodes and the software needed to communicate with them represented a significant development effort. The host computers were ready and waiting; and the telephone lines would be provided by AT&T.[8]

## IMPs Take on the Work of Networking

Roberts awarded the contract for supplying the IMPs to a Massachusetts' company, the Bolt, Beranak and Newman Corporation (BBN). BBN in turn worked with the Honeywell Corporation to adapt Honeywell's H-516 minicomputer to handle the requirements of the switching nodes.[9] Since the success of the network depended on making as few changes as possible to the host computers, the bulk of the technology of the network had to be packaged into the IMP nodes. The IMP had to function as both an interface between the network and the host computers and a packet switch routing the data to its destination.[10] In many respects, the IMPs would embody the network itself. The following list describes the various functions that the IMPs would be engineered and built to perform:

- As an interface, the IMP would take the data coming from the host and translate this data into the packet format used by the subnet. This meant that the IMP would divide up the incoming data into packets of about 1000 bits each (a bit is a binary digit, so think of 1000 zeros and ones) and add descriptive header information into each packet. This header information would include the addresses of the sender and destination as well as a recording of the packet's exact size so that a check could be made for transmission errors. On the receiving end, the IMP would reassemble the packets back into the original data, which meant stripping off the header information and arranging each packet's data back into the original data's order and composition. It would then transmit the reassembled data to the destination host.

- As a packet switch, the IMP would ensure the reliability of the data being transmitted from node to node by executing a *checksum* on each packet: this meant testing the size of the packet against the size recorded in the header information in order to check the basic integrity of the packet. In other words, this process of first recording and later checking the exact size of each packet would help ensure that the data had not been tampered with or otherwise damaged in transit. Once this check was completed and the packet was validated, the receiving IMP would return an acknowledgment to the sending IMP indicating that the transmission was successful. If the sending IMP did not receive an acknowledgment within a given amount of time, it would resend the packet.

■ Also as a packet switch, the IMP would determine independently which route to use to send the incoming packets onwards towards their destination. In order to achieve the military objective of creating a survivable communications network (the military was paying for all this research and development), network routing would need to be distributed, not centralized, as well as adaptable. The IMPs, therefore, would not just know about their nearest, neighboring IMPs on the network and their own, locally-attached host computer. Instead, each IMP would know about the entire network, storing information on every host and on how long and through which IMPs it would take a packet to reach each host. This information would be updated continuously so that changes in network traffic could be factored in and the shortest routes would be recalculated accordingly. Moreover, each IMP would share this data with neighboring IMPs, further helping the whole network to adapt quickly to changing conditions.

The IMPs were carefully engineered to require as little human intervention as possible. In order to meet the government's military objectives and, more immediately, to avoid a maintenance nightmare, the network needed to be autonomous, self-monitoring, self-correcting, and adaptable. Even before the initial four-node network first began transmitting data late in 1969, Roberts was already planning an expansion to fifteen nodes. This expectation of growth, along with the need for the network's independent operation and adaptability, resulted in the development of the following, specific requirements for the IMPs:[11]

■ An IMP would function independently, not relying on other IMPs or on its local host computer.

- An IMP would actively check the network for failures, seeking out lost or duplicate packets, dead lines, malfunctioning IMPs or hosts, and unreachable destinations.

- IMP hardware would be *ruggedized*: a military supplier's term for enhancing protection of components against drastic temperature fluctuations, vibration, radio interference, and power surges.

- An IMP would be capable of being monitored and controlled remotely, enabling centralized operators to run diagnostics or reload the operating system without leaving their desks.

- An IMP would automatically restart in the event of a power failure.

- An IMP would routinely run its own diagnostics and, if a problem was uncovered, would request a neighboring IMP to send it new software to replace the corrupted version.

- An IMP that found a problem but could not repair itself would shut itself down in order to protect the network.

The IMPs did not survive the demise of the ARPANET. But they, more than any other single device, brought the ARPANET to life and handled the bulk of the work of running the network. Much of their engineering, however, lives on today as part of the hardware and software of the Internet's routers and as part of the networking components commonly found in all of our computers.

## The ARPANET Ushers in the Era of Networking

In addition to the technical obstacles, Roberts faced an indifferent, and sometimes hostile, audience in the people who would be attaching their equipment to and using the ARPANET. Not long after the network was up and running, however, most people changed their opinion and hailed the ARPANET as a major achievement for sharing research and facilitating work. Nevertheless, prior to its activation, there was a great deal of skepticism regarding the functioning of the network, concern over the amount of work that would be required to take existing computers onto the network, and suspicion that joining the network would make it more difficult to acquire additional computing resources in the future. Fortunately, Roberts had the upper hand: connection to the ARPANET was mandatory for all computer centers funded through the IPTO.[12]

On November 29th, 1969, the first two nodes of the ARPANET — the University of California at Los Angeles (UCLA) and the Stanford Research Institute (SRI) — exchanged their first packet-switched message. A programmer sitting in the computer center at UCLA attempted to log on to the computer at SRI. But when he entered the *g* in *login*, the SRI system crashed. The local engineers located and corrected the problem later that day. This unanticipated failure, however, was simply the first of many to follow. As with most cutting-edge computer ventures, unanticipated problems were considered commonplace. The ARPANET was an extreme example of such ventures. It represented undiscovered territory, where entirely new computer hardware and software were first interacting with networking theory. No one knew what to expect, other than that problems would occur and resources would be needed in order to find, debug, and resolve these problems.

By the end of 1969, the initial four-node network became operational when the University of Santa Barbara and the University of Utah were added. Within a year, the network was growing at a rate of one new node per month, which was as fast as Honeywell could produce the IMPs.[13] Less than two years after the first data was transmitted across the ARPANET, the planned fifteen-node network was in place and operational.

In 1971, in order to expand access to the ARPANET and its resources, a new type of IMP — called a Terminal IMP (TIP) — was introduced. The TIPs provided direct access to the ARPANET through inexpensive computer terminals. TIPs allowed locations that couldn't afford one of the more costly, IMP-connected host computers to connect to the ARPANET and access its wide variety of resources. As anticipated, the TIPs allowed the network to grow at a much swifter pace and to cover a far larger, more diverse geographic area. Barely two years later, in 1973, half the sites accessing the ARPANET were connected to it through TIPs.[14]

## Protocols Define the ARPANET and its Services

A year before the first IMPs were installed, something called the Network Working Group (NWG) was created, with programmers from each of the four original ARPANET sites. Its task was to determine how the computers on the network would interact and, more specifically, how to create networking protocols that would define those interactions in preparation for the protocols being implemented on each computer. This was a particularly difficult task, not only because the host computers to be attached to the network contained no programming instructions on how to transfer data, but also because these computers knew nothing about the existence of other computers.

A simple function like transferring a file from one computer to another across the network required a protocol: highly detailed rules that specified how the sending computer would open a connection to the destination computer, listen for an appropriate response, and then begin transmitting the data; and how the receiving computer would answer a request, signal a response, and accept the data. This particular protocol — the File Transfer Protocol (FTP) — was the second protocol created by the NWG. The first was TELNET, which specified how you could log on to a remote computer, thereby addressing one of Licklider's and Taylor's initial requirements related to time-sharing.

In addition to specific, application-level protocols, like FTP and TELNET, that defined the networking operations related directly to the tasks we perform at a computer, Larry Roberts wanted the NWG to create a lower, network-level protocol that defined how packets of data would be carried from destination to destination, regardless of the type of data those packets contained or the operation being performed (e.g., a file transfer or a remote login). The NWG created the Network Control Protocol (NCP) to satisfy this requirement.[15] Just as the IMPs embodied the network from the standpoint of the network's hardware, NCP embodied the software side of the network by operating the basic network signals, telling data to wait or go, and so on. In general, every discrete service on the network would require a corresponding protocol. For example, almost exactly twenty years after the ARPANET took computer networking from theory to reality, the World Wide Web would come into existence via its first protocol, the HyperText Transfer Protocol (HTTP).

While these first networking protocols were being written, Steve Crocker, a member of the NWG, distributed a memo with the plain, seemingly inconsequential title of "Request For Comments." In the memo, Crocker documented work being done by the NWG, derived from notes taken during meetings of the group. But he also used the memo to solicit help from anyone and everyone who knew something about the subject matter described in the memo. Crocker's memo, however, resulted in more than feedback and suggestions relating to the NWG's work. It began a process of documentation and debate that continues unabated today and which took on the name of his memo, Request For Comments (RFC).

The RFC process has, among other things, produced a technical library comprising a numbered series of several thousand RFC documents. These RFCs record both the history of the ARPANET's and Internet's engineering decisions and overall architecture as well as the formal, technical specifications of their technology, as exemplified by the RFCs that define the networking protocols. Not surprisingly, RFC number 1 documented the work that had been completed on the host protocols. Eighteen years and 999 RFCs later, in 1987, RFC 1000, entitled, "The Request For Comments Reference Guide", was written.[16] Over the years, not only did RFCs become instrumental in developing the technology for the ARPANET and subsequent computer-based technologies, they were also eagerly adopted for use by many of the computer standards organizations that followed. In many respects, RFCs embody the spirit of the ARPANET, the Internet, and the Web, and the sense of community and inclusiveness that these revolutionary inventions represent.

## Email Turns a Network of Computers into a Network of People

The most significant contribution to network growth, and to the future relevance of networking in the lives of most people, took place in 1972. Electronic mail (called email), first referred to as *net notes* or simply as *mail*, already existed on many computers at the time. These email programs, however, were understandably limited to the exchange of mail between individuals with accounts on the same machine. Then, a programmer from BBN, named Ray Tomlinson, adapted existing file transfer programs to handle the transferring of mail files between machines across the network. He also enhanced an existing program for reading and composing email to accept a new form of email address, one that included an entirely new requirement for the still very young environment of networked computers: the name of the host computer where the email recipient had their account. Tomlinson's email address format consisted of an account name and a computer name separated by an *at* sign (@):

```
user-account-name@host-computer-name
```

News of Tomlinson's innovative work spread rapidly; and it wasn't long before email started to flow across the ARPANET.

Tomlinson's selection of the @ symbol as the delimiter that would separate the user name and computer name portions of the email address was due to the simple fact that it was not then part of any existing user or host name on the ARPANET. Additionally, the effect that an email address read as your name *at* some computer appealed to him.[17] The Queen of England, incidentally, could count herself as a very early participant in the new communication medium of email. She sent out her first email in March of 1976.

The early and ever increasing success of email was not foreseen by the ARPANET's project leaders. But its growth changed the very nature of the network and of the Internet that followed. Within a few short years of the introduction of email, the bulk of the traffic over the ARPANET consisted of email, and email remains the number one use of today's Internet across all ages, groups, and categories of users. The year following Tomlinson's first networked email message, the FTP protocol was officially enhanced to handle the exchange of email; but it would be nearly another ten years before email was defined in its own, separate protocol. The full history of email, as well as the mechanics of how email works, will be presented in chapter 6.

## The ARPANET Flaunts its Stuff

Also in 1972, as ARPA officially became DARPA, the first formal demonstration of the ARPANET occurred at the first International Conference on Computer Communication (ICCC). The infrastructure was in place and the network was operational, but now the immediate problem was that the ARPANET was largely wanting for traffic. People could ship files back and forth, access remote computers, send email and perform assorted other activities, but no resources were available online. For the most part, the network and all its potential were going unused.

The conference, held in Washington, D.C., changed all this by successfully demonstrating the unlimited potential of the ARPANET. Conference demonstrators, consisting of ARPA engineers and faculty and students from the university sites funded by ARPA, impressed the attendees by simulating a distributed air traffic control system that showed a plane departing the air space of one computer's region and automatically being picked up by the computer covering the adjacent region, controlling the movements of a robotic turtle through a networked terminal connection as it made its way around the conference room, connecting to a computer in Paris, testing out meteorological models, demonstrating a virtual conferencing system and a system for displaying Chinese characters, and allowing attendees to test their chess skills against a computerized chess player.[18]

To highlight both the seamlessness of the network and its inherent power for resource sharing, ARPA engineers conducted the following demonstration: sitting in front of a terminal at the conference in D.C., an ARPA engineer logged in over the ARPANET to one of the computers at BBN in Massachusetts, picked up a file of program code from that site and used one of the file transfer programs to copy the file across the continent to a computer at the University of California at Los Angeles, ran the program code on that computer but directed the program's output to a printer adjacent to where the demonstration was being held at the conference.[19] Such utter disregard for where a resource was located or where information was stored in order to perform a task was new and startling. It's very likely that everyone who witnessed this demonstration walked away thinking about how they might apply this new paradigm of computer use in their own environment.

The conference marked a turning point. Suddenly, the experimental ARPANET, a testbed for networking theory and technology, had been transformed into a functional tool with exciting and practical applications. In the month following the conference, traffic increased 67%; and that was just the beginning. The ARPANET was something new and people needed to see it for themselves in order to grasp its significance. With increased popularity came greater change for the ARPANET. But it was not the ARPANET itself that inspired and motivated change. It was the more general concept and application of networking that started to change the landscape of computing and, with it, change the ARPANET, too.

By the time of the 1972 demonstration, the ARPANET had succeeded in meeting its functional requirements: the original fifteen IPTO sites (and more) were connected, data could be easily transferred from host computer to host computer, computers could be accessed remotely, and packet switching had proven itself and had provided the military with a survivable communications network. But the ideal of resource sharing never materialized; nor did the notion of distributed computing. Getting started with accessing and using the ARPANET, and locating the resources that were available over it, was a challenge for people who were not computer experts. Roberts and the IPTO could mandate that all of the computer resources paid for by the IPTO at the ARPANET sites be freely accessible and available for one and all to use across the network. But they could not mandate how, or if, people would use those resources.

What instead took place was that people started to apply their newfound appreciation for networking in highly original and unanticipated ways that suited their own needs. The arrival and popularity of local area networks, which are presented next, is one example of this behavior. This was the beginning of a trend that had a great impact on the remainder of the ARPANET's twenty years of operation. Moreover, this same innovative spirit and grassroots approach to networking was also largely responsible for making the Internet into what it is today; and it continues to shape the Internet's form, functionality, and technology right now, no doubt affecting what it will evolve into tomorrow.

## Local Area Networks Arrive Spontaneously

In the early 1970s there was no such thing as a local area network (LAN) through which computers located in the same room or building could communicate with one another. Users at connected ARPANET sites could easily ship files across the U.S. But to transfer a file between two computers in the same room they would have to copy the file onto magnetic tape and then copy it from the tape onto the other computer. Creative ARPANET users at MIT, however, figured out a way to turn the IMP that connected them to the ARPANET into the hub of a LAN. Their IMP now connected them to the ARPANET's wide area network (a network of computers distributed across any distance), with all its remote hosts and resources, while it also interconnected their local computers, thereby enabling the same file and resource sharing in their local environment. Network monitors at BBN quickly noticed the increase in traffic at MIT's IMP. But they did not know what to make of the fact that there was no corresponding increase in traffic over MIT's outgoing lines.[20]

Such a use for an IMP had simply never been considered. But when other sites learned what MIT had done, using an IMP to create a LAN quickly became a common practice. By 1975, nearly one-third of all ARPANET traffic was LAN-based: local, intra-node traffic among computers at a single location. This was quite a shock to the ARPANET management. Use of the ARPANET as a LAN continued well into the 1980s, at which point Ethernet (a simpler, less costly, and more efficient LAN technology) quickly became the de facto standard for establishing a LAN, eliminating the need to use an IMP as a local network hub. But this simple, imaginative, purposeful innovation added substantially to the usefulness of the ARPANET. It was among the first of many such innovations that were created by users of the network, then swept across the network, and ended up changing the network's very nature.

## Packet Radio and Satellite Networks Emerge

Another innovation of the time occurred in Hawaii and was called the ALOHANET. Funded by Roberts through the IPTO, as well as by the Navy, Norman Abramson developed an ARPANET-like network using radio signals to connect seven campuses at the University of Hawaii, as well as Hawaiian research institutes, with the university's computer center near Honolulu. A radio-based packet-switched communication system, referred to as a packet radio network, was needed for Hawaii because the existing telephone lines were too noisy for reliable data transmission. Without a reliable telephone network, the ARPANET's IMPs and their packet-switched network could not provide the necessary speed and efficiency for transferring data across the network. In addition, a working packet radio network would be perfect for battlefield military communications, which already relied on radio-based communication; and showing a direct correlation between a funded project and its military application was always a key concern of ARPA and the IPTO.

As will be described in chapter 4, a packet radio network posed a wide range of new engineering challenges, not the least of which was how to deal with the collision of two or more radio signals broadcasting at the same time.[21] Much of the work Abramson and his team did to resolve these problems led directly or indirectly to the wireless communication technology commonly used today, including cell phones and portable internet terminals.[22]

Another packet radio network — called the PRNET — went into operation in the San Francisco Bay area in 1975 under the guidance of Robert Kahn at ARPA. Kahn wanted to use the PRNET to explore voice communication over packet radio. It was thought that voice transmissions over a packet radio network would be more efficient than the existing radio channels used for short range communication. Also, it could be configured to correct for noise-induced errors. Most importantly, packet radio voice transmissions would be far more difficult to intercept and eavesdrop on since messages would be converted into a digital format, divided into packets, and have to be reassembled and decoded on the receiving end before they would make any sense. Such a clear military application made the PRNET, like the

ALOHANET, a valuable project in helping to convince the Department of Defense about the merits of conducting such research.[23]

Another experimental network of the time was ARPA's packet satellite project, the SATNET. The benefits of using satellites were that they offered higher bandwidth (they could transmit more data in the same amount of time than other transmission architectures, like the IMPs and the telephone cables that interconnected them) and they could cover a wider area, providing, for example, a data bridge across the Atlantic or Pacific. They were, however, rarely used for transmitting data because the cost was prohibitive. But the inherent economies of packet switching suddenly made using satellites much more feasible.[24]

The SATNET was sponsored by ARPA, the British Post Office, and the Norwegian Telecommunications Authority. It was created to support network research and, more specifically, to transmit seismic data that was generated by sensors in Scandinavia and needed to be transferred to the U.S. for analysis. Initially, the SATNET connected four sites: one in Maryland, another in West Virginia, one in England, and one in Norway. The SATNET was a hybrid system. Like the PRNET, it was a broadcast system that used radio channels to transfer data between the stations and the satellite. Like the ARPANET, it used packet switches similar to the IMPs, but modified to handle the higher bandwidth and the longer transmission delays.

## The Internet Project

As Kahn considered developing the PRNET beyond its experimental phase, he realized that its future lay in connecting it to the resources of the ARPANET. It quickly occurred to him that the same was true for the SATNET. But connecting these networks presented considerable challenges. All three networks were essentially incompatible. For instance, while the PRNET broadcast its packets and did not ensure delivery (it transmitted its data without specifying a destination; the data would be received by any one of several receiving stations), the ARPANET transmitted its packets from one point to the next and guaranteed that the transmission would be reliable and that the packets would be sent in sequence. In addition, the size of the packets and the transmission speed differed among the three networks. In a small, but critical way, Kahn's effort to connect up these dissimilar networks marks the beginning of the Internet, a network of independent, interconnected networks.[25]

A paper written in the summer of 1973 by Kahn and Vinton Cerf (one of the designers of the ARPANET's host protocol, NCP) served as the starting point for the architecture of the ARPA Internet. In it, they attempted to determine what kind of host protocol was needed to compensate for the error-prone nature of the packet radio network; and they explored how two separate networks, very different in their engineering and functioning, like the PRNET and the ARPANET, could be connected. Kahn and Cerf then went in search of help, soliciting comments and assistance from networking experts in the U.S. and overseas. The broad range of help they received was critical to the long-term success of their work. For one thing, they were encouraged to look beyond the PRNET's packet radio network and its particular issues and examine as well the different types of packet switching networks that had been developed in Europe.[26]

The first part of their solution came in the form of a new host protocol called the Transmission Control Protocol (TCP). It was intended to replace the existing ARPANET host protocol, NCP. TCP was a significant departure from NCP, which was designed under the assumption that the network, specifically the subnet embodied by the IMPs, would be totally reliable and error free. Accordingly, NCP did not even include any error recovery procedures. The IMPs, not NCP, handled issues related to data and transmission reliability. TCP, however, would effectively turn the equation around: the network would be assumed unreliable; and the hosts, not the network, would take full responsibility for the reliability of the data.

TCP incorporated the best features of two emerging networking technologies of the time. The first was the simplified packet switching system of the Cyclades project in France. Cyclades, named after a group of islands in the Aegean Sea, was designed with interconnecting networks in mind and shifted most of the requirements for packet switching from the network, where the ARPANET had placed them, to the hosts. Cyclades extended the design of packets with something called a *datagram*. With

Cyclades, the network simply passed datagrams from node to node, while the hosts took responsibility for controlling flow, handling errors, and sequencing.[27]

TCP also took inspiration and engineering from Ethernet, a local area network technology developed by Robert Metcalfe in the early 1970s that, from the 1980s on, has been a de facto networking standard. Metcalfe, in turn, was inspired by the ALOHANET and applied many of the principles of that network to a network employing a cable, in place of radio signals, to interconnect computers. His engineering focused on transmitting packets over a short distance as quickly and efficiently as possible. With Ethernet, there is no intelligence in the network itself, just packets being sent and received by hosts along a length of cable. In Ethernet's model of networking, much like Cyclades', the hosts become the network and each computer host is equal to every other host on the network. It's interesting to note that the first version of TCP seriously underestimated the potential growth of networking. The original design could only accommodate a maximum of 256 networks. By the late 1990s, however, the number of networks on the Internet had surpassed 100,000.[28]

The second part of Kahn and Cerf's solution was the creation of network gateways. A gateway is a specialized host computer that connects two or more networks. Each gateway would be responsible for maintaining routing tables in order to direct packets between the networks and for translating packet formats when necessary to accommodate differences in each networks' packet specifications. Just as the IMPs of the ARPANET facilitated the transmission of data among any number of different types of host computers, the gateways allowed each network to remain indifferent to and removed from all the various network incompatibilities. This design contributed substantially to the scalability of the ARPA Internet since any number of new networks could be added by simply adding a new gateway to interconnect the new network with the larger ARPA Internet.[29]

Along with the concept of gateways came new naming conventions for network addresses. Unlike the ARPANET, where network addresses were not even needed since packets were sent to a specific IMP which in turn was attached to a single host, the Internet required each computer to have an address that indicated both its name and the name of its network. Accordingly, ARPA engineers designed a hierarchical address system that allowed for as much or as little granularity as people wanted to specify.

Just like a telephone number or a postal address, each computer on the network needed a unique identification in order for data to reach its intended destination. The simplest computer address would contain only a host name followed by the name of the network to which it was connected, such as `frodo.mit`. But any number of domains (discrete network subdivisions) could be specified in an address, up to what was called a fully qualified address, one that included all domains in the network hierarchy, such as:

```
frodo.theshire.students.engineering.mit
```

This system became known as the Domain Name System (DNS) and helped to simplify both network management and the routing of data.

The format of a DNS network address worked much like the format of a postal system's address. Each portion of the address helped to refine the route and pass the data further along towards its destination. Just like the postal system with its address refinement from the most general to the most specific — from country or state, to city, and finally to street and house number - DNS accomplished the same task through its specification of domain names. In the example above, the "mit" domain would be the most general, and from there data would route through each of the subordinate domains until it reached "theshire" network to which the computer named "frodo" was connected.

Overall, DNS was divided into two functional parts: one part stored all the host and network names alongside their location information (a numeric identification known as an IP address) in a hierarchical database distributed throughout the network (think of an online phonebook); the other part consisted of programs to query the database in order to retrieve a computer's location by supplying its name (think of using someone's name to look up their phone number or address in the phonebook). In this way, DNS established a globally unique address space, a way to identify and locate computers and computer networks throughout the world. More information on how DNS works, and the significance of DNS in defining the Internet, will be presented in chapter 3.

The gateways, in conjunction with DNS, helped manage the complexity of Kahn and Cerf's design for the Internet. The gateways would route packets from network to network, without needing to know how those networks were subdivided or where the individual hosts were located. Similarly, hosts did not need to know about how the local network was organized or about anything outside their domain, while nodes on the local network did not need to know about non-local hosts or routing but only how to get packets to the gateway.[30]

One initial design flaw was that gateways added a layer of complexity to the network since they used TCP to route the packets. In so doing, they duplicated the same error control and sequencing functions performed by the hosts. This problem, and others that emerged soon afterwards, caused Kahn and Cerf to divide TCP into two separate protocols in 1978, as follows:

- A host-to-host protocol (TCP) to order packets, provide verification by issuing acknowledgments, perform error checking by requesting retransmission of lost or damaged packets, and control traffic flow by limiting the number of packets in transit.

- An internet (initially called, internetwork) protocol (IP) simply to pass packets, either between a host and a switch (an IMP on the ARPANET, or some other network device capable of handling the routing of data on a packet-switched network) or between switches, which included gateways.

These protocols became collectively known as TCP/IP. The creation of IP allowed gateways to be simplified and stripped the requirements for network traffic down to a bare minimum. This had the added benefits of increasing the robustness of military networks and allowing for the creation of new, highly specialized networks, without disturbing the existing system.[31] Chapter 5 will present more detailed information on the Internet protocols, including a high level, working description of TCP/IP.

## The Internetworking Demonstration

In 1977, to show the U.S. Department of Defense what they could do with these newly interconnected networks, DARPA conducted an Internet experiment in which data packets were sent on a 94,000 mile round trip across several different networks. The data started its journey from a van traveling across the Golden Gate Bridge in San Francisco via the packet radio network of the PRNET. From the PRNET it went to an ARPANET gateway and through it, via a series of ARPANET TIPs and gateways, to Norway and on to London. It traveled through the packet satellite network of SATNET to an IMP in West Virginia, where it continued via the ARPANET through Massachusetts and finally returned to California.[32]

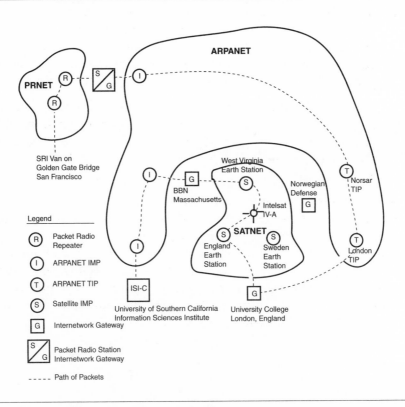

**Figure 1. 1977 Internet Demonstation**

The system worked flawlessly and was hailed as a complete success. No data was lost. Even while crossing the bridge, the radio signal transmitted correctly and automatically retransmitted any packets lost or damaged by the huge, steel structure. This first operational Internet experiment clearly demonstrated that individual, local networks could remain autonomous and control their own traffic and development while maintaining a connection to other networks. As a result, several defense and research networks joined the ARPA Internet shortly following the demonstration, thereby increasing the size and usefulness of the network as a whole.

## The Great Protocol Migration: TCP/IP or Else

During the next few years, from the late 1970s through the early 1980s, ARPANET sites were encouraged to migrate to TCP. Meanwhile, ARPA funded implementations of TCP for IBM and DEC machines, as well as for Unix and other operating systems. But the transition was technically challenging, and most sites continued to use the old protocol, NCP, for as long as possible. Ultimately, the U.S. military mandated the switch and a cutover date of January 1, 1983 was set.

Military interest in and control over the ARPANET project had grown in intensity as the experimental nature of ARPA's research and development efforts became less tenuous and as the practicality and usefulness of the system expanded. In 1975, the Defense Communications Agency (DCA) took over official operation of the ARPANET and began using it for its new, computerized command and control systems for the armed forces. There were at the time some 66 nodes and between 4,000 and 5,000 users.[33] The informal, little policed, early ARPANET policies started to be replaced by a formal bureaucracy that set out to enforce the network access policy and prohibit all unauthorized use.

Concerns about security on the ARPANET increased dramatically with the introduction of the very first personal computer into U.S. markets in January, 1975. The computer was the Altair 8800, and it sold for $379. This small, simple computer, with barely any computational power, represented a threat far greater than its size or price. It meant that access to computer technology would soon be available to a new and far wider audience. Concerns were raised that the availability of this technology would produce a *hacker* subculture and represented a security threat to existing computer systems accessible on the network and to the network itself. The government cited justification for these concerns in, for example, the penetration of telephone system security by means of something called *blue boxes*, simple electronic devices that mimicked the telephone's control tones to fraudulently acquire free telephone service.[34]

The migration from NCP to TCP/IP in 1982 and 1983 was traumatic, according to most accounts. The magnitude of the work and the harsh, well-communicated threat of being cut off from the network if one failed to migrate, make the late 1990s Y2K preparations and anticipation look trivial by comparison. In order to migrate to TCP, each networked site had to perform the following work:

- Write/acquire TCP software for each host computer.

- Adapt existing FTP and TELNET software to meet the new, updated protocol specifications.

- Adopt the new mail standards: a new protocol called Simple Mail Transfer Protocol (SMTP) and a new mail addressing scheme.

- Replace existing IMPs and TIPs with new devices that ran IP.

Only half the sites met the January 1, 1983 deadline. Many sites, not having taken the deadline seriously, were surprised when their network access was terminated. Others found errors in their TCP implementation. Provisions were made to extend the deadline on a site by site basis, but only after paperwork was submitted to request a temporary exception. By March, half of the remaining sites had successfully made the transition; by June, all hosts were running TCP/IP.[35]

## Success and Separation

With the migration to TCP/IP complete, one major obstacle to the establishment of today's Internet had been overcome. Another event in 1983, a more direct response to the security threats described above, was the division of the ARPANET network into two networks: one, called the MILNET, for the military; the other, which continued functioning under the ARPANET name, for scientific research. In this way, military communication could operate under a more controlled and restrictive system, one that included encryption devices and additional security apparatus; and the ARPANET could return to focusing on network technology development and continue to grow and flourish in an environment dominated by research institutions and universities, making it that much easier to transfer its authority one day to civilian control.

Two key strategies facilitated the creation and successful development of the ARPANET, and they continue to be responsible for the ongoing development of the Internet. One, a development methodology called *layering*, took complex tasks and divided them up into discrete, modular components. These discrete components were often arranged in layers, starting with the simplest, most concrete units and building, layer upon layer, to the most abstract and complex units. The other strategy consisted of an informal, decentralized management style in which participation was welcomed and often solicited, as exemplified by the RFC documents.[36] Both strategies will be discussed in detail in chapter 8.

# Shutting Down the ARPANET

Starting in 1988 and finishing in 1990, after twenty years of breaking new ground and changing the way scientists, governments, and individuals accessed and employed computers, the ARPANET was shut down. The shutdown process was yet another story of migration in the history of the ARPANET, as each ARPA site switched its host connections from the ARPANET backbone to a newer, faster backbone run by the National Science Foundation (NSF), called the NSFNET. The NSF, its network, and the other networks that started up and thrived during the 1980s will be discussed at length in the next chapter.

Networking technology and network usage grew during the 1980s at a rate no one could have predicted. LANs sprouted everywhere, expanded quickly, and started interconnecting to form still larger networks. Single-user personal computers changed both work and home environments, while research and academic institutions adopted workstation minicomputers running the Unix operating system, most of which came ready to be plugged into an awaiting network. A mere 2000 computers had Internet access in the fall of 1985, while just two years later nearly 30,000 had access; and by October, 1989 this number increased more than fivefold to 159,000.[37] The following chart nicely summarizes some of the ARPANET's key developments and events alongside its growth in terms of traffic and number of hosts:[38]

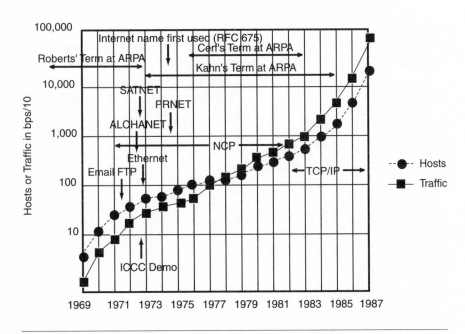

**Figure 2. ARPANET/Internet Growth and Developments**

More computers meant more users, which in turn meant ever-increasing traffic across the networks. The ARPANET provided the backbone across which other networks transmitted much of their data. By the end of 1987, however, ARPA's managers determined that the ARPANET's architecture could not keep pace with the growing demand. Its IMPs were not designed to handle such capacity and the lines connecting them were equally deficient, operating at only 56 Kbps. Since ARPA and the NSF were already working together, many sites already connected to both networks. For these reasons, and the others sited above, the NSFNET was an easy and obvious choice to replace the ARPANET.[39]

The ultimate tribute to the builders of the ARPANET can be found in the fact that the changeover and decommissioning of the ARPANET went largely unnoticed. Few service disruptions took place during the transition period; and the vast majority of Internet users were not even aware that anything had occurred.[40] More importantly, the technology ARPA's scientists and engineers designed, tested, and implemented during the twenty-year evolution of the ARPANET survives today in the form of the Internet, and in myriad other private and public networks, and provides the foundation for much of the networking technology we use in the 21st century. While some of their original goals never materialized, what they created turned out to be something larger and broader in scope than what they had originally envisioned. The lasting effects of their work can be seen today in the continuing advances found in new networking products, a broad range of new tools through which people can interact and communicate, the new ways for businesses to promote and sell products, and a whole host of other wonderful inventions and innovations, many of which will be presented later in this book.

# The Internet: A Network of Networks

3

## A Slow and Quiet Beginning

Unlike the ARPANET, no one set out to engineer and build the Internet. Nor was the Internet invented. Furthermore, the Internet represents no individual's, nor any one organization's, master plan for a global network. This is partly why no person, group, organization, or country can claim ownership of the Internet. This is also why no single, all-powerful controlling body governs the Internet's operation, determines what services it will and will not provide, or decides who can and who cannot use its resources.

The lack of any planned direction, demonstrable ownership, or authoritative control are features that distinguish the Internet from other technologies, from other forms of communication, from other information access systems, and, more generally, from other pivotal developments — technology-based or otherwise — that have directly and dramatically effected changes to us as individuals and to the way we live and interact with others. As will be explained below, these same features also contribute to the uniqueness of the Internet, to the strikingly organic nature of its existence, and to its capacity for growth and change.

The Internet acquired its unique form and its various component functions and features slowly and quietly, through subtle transitions in its control and oversight, and through seemingly innocuous and simple innovations that, in the end, showed themselves to be anything but simple or small (think of email and the World Wide Web). At first, the Internet went largely unnoticed by most individuals and organizations. Over time, however, the Internet expanded its influence, until, eventually, it achieved prominence through a type of passive, grassroots evolution of its technology and a corresponding expansion of, and appreciation for, its various services and information sources. When, finally, the Internet acquired the full scrutiny of myriad government agencies, large and small commercial concerns, and an assortment of local, national, and international organizations, all clambering to control the Internet's direction in one way or another, the Internet had already firmly established its existence as an entity collectively owned, operated, and controlled by the distributed and wide assortment of individuals and groups responsible for its services and for its ever growing use and popularity. Even the rich and powerful telecommunications industry failed in its attempt to exert control over the Internet. The failure of its effort, much like the failure of other groups, can be generally attributed to a lateness in its timing, a noticeable overconfidence in its approach, and an utter disregard for the Internet's intrinsic environment with its powerful and tightly bound community structure.

The Internet's earliest beginnings trace back to ARPA's original Internet Project and the considerable energy and engineering that went into interconnecting the ARPANET with two other independent networks: the satellite network of SATNET and the packet radio network of PRNET. Moreover, the Internet, in its earliest forms as well as today, owes its existence to the engineering and evolution of TCP/IP. The protocols of TCP/IP effectively form the *net* of the Internet; and, more than any other single creation, they are responsible for the Internet's rise to dominance and for its growing and pervasive popularity.

The Internet that we know today, however, didn't really come into existence until the government-owned-and-operated, highly restrictive network environment developed and administered by ARPA began to intersect and interconnect with the commercial, freely accessible networks that were the result of the many academic and commercial forays into the new environment of networking and networked computer systems. The convergence of all these different networking enterprises and the organizations behind them, which occurred near the end of the 1980s and into the early 1990s, is what gave form to the Internet we know today.

The U.S. government, as well as several European governments, played an essential role in transforming the Internet architecture from concept to experiment to thriving operation. The private sector also played its part, first by adopting the architecture and integrating it into more and more computer systems, and later by introducing the Internet to the public, offering services and products for new sorts of commerce, personal communication, gaming, and more. The creation of workstations, the popularity of the Unix operating system, and the emergence of local area networks through the wide acceptance of Ethernet's simple and efficient networking technology, combined to help establish and nurture the Internet industry, as members of the research community saw the promise of the Internet and set out to start or join the dot com revolution. The introduction of the personal computer accelerated the growth of interest in and access to the Internet, bringing computers into the home and taking homes and individuals onto the Internet.

The story of the Internet is less a history of dates and people and events than it is a story of small, but far-reaching technological innovations and the effect of these innovations on the ways in which we communicate, acquire information, and interact. It's the story of an evolving architecture that grew out of the scientific and military environment of ARPA and the ARPANET. But it's also the story of an equally innovative, but far more diverse and inclusive environment of early computer networking enthusiasts who introduced networking to the established business community and to the home computer user. Along the way, computer networking slowly but surely became another venture in the interests of big business. Corporations discovered how networking could reduce their costs and increase the efficiency of their operations, and how the Internet could be employed to create entirely new business enterprises or revitalize an existing business. The result was the addition of yet another dimension to the Internet's composition, use, and impact. Eventually, as business, government, and the commercial and personal interests of individuals converged, the final obstacles separating the old Internet from today's Internet were removed: the U.S. government stepped back and privatization of the Internet took control; and the prohibition against commercial use was removed, allowing commercial traffic to start flowing freely across the Internet.

This chapter presents an historical perspective on how all of this happened. It describes what constitutes the core architecture of the Internet and how this architecture evolved. It also presents a look at some of the private networks that emerged while the government-controlled ARPANET and early Internet were being developed. It then relates the history of the control and oversight of the Internet, its transition to a private, commercial operation, the short-lived and spurious battle for control of its technology, and the arrival of the World Wide Web and its effect. It also explains how the daily operations of the Internet are managed. Finally, it presents a cautionary tale regarding one of the earliest malicious attacks on the Internet, the Internet worm of 1988. First, though, a short, simple definition of the Internet is presented.

# A Working Definition of the Internet

So, what is a *working* definition of the Internet, a sort of baseline description that everyone can agree on? The shortest, simplest definition may also be one of the earliest. It describes the Internet as a network of networks, or an internetwork of networked computers. This definition conveys a sense of the Internet's overall function and the role it serves as a type of master, unifying network. But it does not help to explain what makes the Internet distinct, how it works, or even what constitutes its basic structure.

Any working definition of the Internet must identify what it consists of in order for us to determine what is, and what is not, the Internet. Because the Internet has permeated so many areas of our personal lives, commerce, the legal system, and government, more formal definitions have been written to help clarify and qualify the Internet. A good example is the definition created on October 24, 1995 as a Resolution of the U.S. Federal Networking Council:

The Federal Networking Council (FNC) agrees that the following language reflects our definition of the term "Internet."

"Internet" refers to the global information system that --

(i) is logically linked together by a globally unique address space based on the Internet Protocol (IP) or its subsequent extensions/follow-ons;

(ii) is able to support communications using the Transmission Control Protocol/Internet Protocol (TCP/IP) suite or its subsequent extensions/follow-ons, and/or other IP-compatible protocols; and

(iii) provides, uses or makes accessible, either publicly or privately, high level services layered on the communications and related infrastructure described herein.[1]

Note how the definition includes both the highly specific (the reference to the significance of TCP/IP) and the very general (the final item and its deliberately vague language about how the Internet is used or accessed). It is hard to imagine the Internet functioning without its core dependence on TCP/IP, the component most responsible for enabling all of our disparate computing devices to interconnect and share information. It is equally hard to imagine what new and highly inventive uses will be made of the Internet, as more and more people start using it and as the world's societies grow to depend on it even more than they do already. It's this very combination of core technology with unrestricted, generic services that made possible the widespread interoperability and use of the Internet, and that best characterizes its composition.

As a working definition, the following should suffice: the Internet is a global information system of interconnected networks that relies on the fundamental communication protocols of TCP/IP.

# The Architecture of the Internet

The Internet began as an architecture: the solution to an engineering project that called for the interconnection of several different types of networks, without any disruption to the operation of those networks and without any changes to their infrastructure. The architecture that resulted from this work, like the definition of the Internet presented above, incorporates elements that are, on the one hand, specific and highly technical, and, on the other hand, general and more conceptual in nature.

This division in the Internet's architecture into the specific and the general correlates to its two main functions. The specific portion relates to the Internet as a communications medium, delivering packets of data and providing end-to-end, or host-to-host, communication services. TCP/IP was created to provide for this part of the Internet's architecture, establishing a common language and common framework that allow each of our computers to connect to the network of the Internet and to send and receive data. The general portion relates to the Internet as an information system, enabling individuals to create, store, and access any and all types of data, and doing so in ways independent of the underlying communications infrastructure. All the services that we interact with when using the Internet, such as email and the Web, constitute this part of the Internet's architecture.[2]

The implementation of the Internet's architecture resulted in the creation of its three core, technology-based components: the networking protocols of TCP/IP, the networking hardware of gateways (now called routers), and the networking address mechanism of the Domain Name System (DNS). Think of TCP/IP as the means by which data travels the Internet: it packages the data for its trip, operates the stop, go, and resend traffic signals along the way, and unpacks and checks the data on the receiving, or destination, end. Over time, TCP/IP became the all important standard that allowed the Internet's basic architecture to permeate all types and brands of computers and, later, smaller, more mobile network-capable devices, such as cellular phones, beepers, and personal digital assistants (PDAs) that send and receive email and access the Web.

Think of gateways as key interchanges, or control points, on the Internet: they provide the essential service of accommodating and simplifying all the differences between the wide variety of networks that exist today; and they facilitate the creation of new networks and their connection to the Internet. It's largely the gateways that keep the Internet running as new technologies are introduced and existing technologies are enhanced or altered. The gateways help to isolate many of these changes, allowing our computers to remain unchanged as the Internet continues to evolve.

Finally, think of DNS as the means to identify and locate computers and their resources on the Internet: it handles the more mundane, but necessary chore of keeping track of the hosts that constitute the Internet, thereby enabling data to find its way to its intended destination. DNS established a hierarchical computer naming, or addressing, scheme that provided each computer with a unique Internet address. It also provided the means to store and look up the addresses of other computers.

If you want to define the Internet by its architecture or its technology, then you don't need to stray far from TCP/IP, gateways, and DNS. But the story of the Internet is more than a story about its technology and overall architecture, as the following sections make clear. More than anything else, it's the story of the creation and development of all types of different networks and the new and different services they offered, something that could not have happened without the creation of the ARPANET and the early Internet. It's also a story of control — both political and economic — over technology, network access, and management.

## The Emergence of Other Networks

While ARPA kept busy with the big picture of networking, providing the highways across the U.S. and the connecting routes overseas to interconnect the U.S. with Europe and elsewhere, a diverse mix of institutions, corporations, and creative individuals fashioned their own assortment of new and unique networks to meet local needs. These networks were responsible for introducing the power and usefulness of network communication to a larger and different audience than that served by the government-controlled and highly restricted ARPANET.

Some individuals saw networking as simply a means to an end: connecting computers to transfer files, logging in remotely to work from home or to gain access to an otherwise inaccessible machine, or backing up data from one machine to another. Others envisioned networking as a facilitator of new products and services, an enabler of otherwise unattainable goals: creating electronic, online versions of community bulletin boards, building online forums for information sharing and problem solving called newsgroups, talking across the network via a window on a computer terminal, or self-publishing documents for others to read.

Such efforts created the networking environment of the 1980s, an environment better suited to the computer-literate and technically minded minority than to the general population. As many of these independent networks grew in popularity, attracting more and more people from an ever widening geographic area, the data they carried came to travel the same backbone networks that carried the traffic of the ARPANET. It was this intertwining of traffic that gave rise to a pre-commercial Internet not all that dissimilar in function to today's. Since, however, the commercialization of the Internet was still several years away, there were as yet no Internet service providers (ISPs) to provide general access. Instead, access to most of these independent networks, and to the larger Internet, was primarily through a government, university, or research institution account. Nevertheless, these early networking efforts paved the way for the more egalitarian and commercial Internet of today. The following sections present a small, representative sampling of the many networks and services that came into existence during this period.

## Bulletin Boards

In the 1980s, bulletin boards served as a kind of precursor to network and Internet access for many individuals with personal computers. By attaching a modem to your computer and installing some basic software, you could dial a telephone number and log in to a free or fee-based bulletin board. The effect was more like gaining access to a larger, remote computer and accessing communal resources than connecting to a network. Imagine connecting to the Web, but being restricted to one site.

Bulletin boards offered access to an early type of online virtual community. Since these bulletin boards typically catered to a specific geographic region, the information they contained, like the people who visited them, were local to one area, further fostering their significance as a new type of social environment for the community. Technically speaking, everything was quite rudimentary by today's standards. This was before browsers and the Web. Even graphical user interfaces were still years away. Displays were limited to showing text, and only text; there were no animations or graphical elements of any sort, only what could be input through a standard keyboard. The information presented on bulletin boards took the form of lists and text menus. You moved your cursor by using arrow keys to scroll up and down; and you made a selection by pressing the enter or return key.

Many bulletin boards focused on discussion groups that allowed people to express opinions and share information. You connected to a bulletin board to read through a chronological listing of comments on a particular subject (often referred to as a thread), to post your own remarks, and to see how your comments were received and decide whether or not to respond. Bulletin boards also served as specialized mail gateways, enabling individuals to exchange both personal and public messages with local bulletin board members and sometimes with members at other bulletin boards. They also served as repositories of software, from which you could download games, software fixes or patches to known problems in games and applications, and other types of files.

## FidoNet

Bulletin boards offered a sense of community and, more often than not, were established by individual computer enthusiasts or hobbyists who wanted to share information by providing a welcoming environment, some basic resources, and a simple set of rules. FidoNet, which began operation in 1983 and is still running today, brought bulletin boards into the realm of computer networking through software that interconnected bulletin boards into their own, independent network. The FidoNet bulletin board system (BBS) software was originally developed by Tom Jennings and was designed to create an electronic mail and conferencing network for hobbyists. The idea was to establish local access to more diverse subject matter by building a far more widespread, but interconnected community of users and a greater wealth of shared information than could be achieved by any single bulletin board. The network was simple in both its design and functionality, consisting of a small collection of home-based computer nodes that called each other over the phone lines to share and synchronize their data.

The growth of FidoNet, like that of the Internet, was fast and furious. Also, just as with the Internet, each new spurt of expansion brought with it new technical obstacles to overcome. As the network grew, simple and commonplace tasks, like handling errors caused by noisy phone lines or misbehaving computers, became more troublesome and far-reaching. Even the initial method for routing data from node to node across the network could not keep pace with the network's exponential growth. As a result, a node hierarchy was created (not too dissimilar to that established by DNS for the Internet) that allowed data to be routed from higher- to lower-level nodes, enabling far better efficiency and making is possible to accommodate future network growth. What started out as an informal club became, and still is, a self-governing organization defined by some basic rules and regulations. FidoNet functions as an independent network; but like many such distinctive and independent networks, it is also interconnected with other networks and with the Internet itself.

## UUCP

In 1976, AT&T Bell Laboratories developed a network file transfer program called the Unix-to-Unix Copy Program (UUCP). As its name indicates, the program was written to copy files from one Unix machine (a computer running the AT&T-created Unix operating system) to another Unix machine. It was distributed freely with the Unix operating system in the following year. The program was a simple, but well designed utility; and it enabled system administrators to easily configure their Unix systems to call other known systems, that were also using UUCP, in order to exchange files. This type of network was labeled a *store-and-forward* network, since it was fundamentally a method for distributing files by copying them from location to location.

At first, the calls were made over existing phone lines, but later the program was enhanced to use a local area network and the Internet. Any type of file could be exchanged; and the systems could be configured to call on demand (start up the connection to a specific machine as soon as a request to copy a file was issued) or to poll other systems at certain times or intervals and send and/or receive waiting files at that time. The first uses of UUCP included the distribution of operating system and program file updates. A somewhat different use of UUCP consisted of remote command execution, which might involve, for instance, submitting a file to a printer attached to someone else's computer or network.

In the days before local area networks, UUCP provided a fast, easy, inexpensive way to establish a limited, but highly useful network. This network could just as easily interconnect computers in the same building as form a network of computers distributed throughout a country or the world. Much like the services of bulletin boards, UUCP quickly became a means of sharing all types of information. But perhaps its most popular use was as a distributor of email. Even later, when local area networking became common, UUCP was still frequently employed to create small, self-contained networks with minimal formality, cost, and regulation.

## USENET

One of the most popular of the UUCP networks was USENET, a distributed network containing bulletin-board-like news articles. It was created by a couple of Duke University students in 1979 with three computers located in North Carolina. Once referred to as the poor man's ARPANET, USENET grew exponentially over the years and remains to this day a large and thriving presence on the Internet. One of the reasons for its continued success is that it has been continually adapted and enhanced in response to changes in computer hardware, new standards in networking, and the creation of new applications and network services. For instance, when TCP/IP became a networking standard, USENET was adapted to work over TCP/IP. Similarly, when the World Wide Web came into existence, viewing and interacting with USENET became possible via a Web browser.

Despite the changes over the years in USENET's underlying technology, or in how its articles are posted or accessed, the fundamental purpose and use of the network has remained unaffected. USENET continues to be a world-wide, distributed network dedicated to the free and uncensored sharing of news and all other forms of information. In many respects, USENET can be considered the Internet's discussion forum. Both the Internet and USENET exist and function thanks to the grassroots efforts of individuals and groups scattered throughout the world. Neither have any centralized control or management; and the one word that may best describe them both is diversity. Witness to this fact is the amazing breadth of information that USENET offers, with tens of thousands of newsgroups — ranging from technical computer questions to humor to cooking — covering just about any subject you can think of, and with contributions from individuals located all over the world.

## BITNET

Another store-and-forward type network, like the ones based on UUCP, was called BITNET (the 'BIT' stands for 'Because-It's-There'). BITNET started operation quietly and modestly in 1981. It used a protocol — native to IBM machines — called Remote Job Entry (RJE) that, like UUCP on Unix machines, made possible the remote execution of programs. IBM funded the original experimental connection between Yale University and the City University of New York; and this became the first such network for IBM users. It was used primarily for the exchange of email, but it also allowed two people to 'chat' over a dial-up connection in real time.

BITNET was to educational institutions and scholars what the ARPANET was to the government-funded scientific community and researchers. It thrived in the 1980s as a means of forming communities of individuals interested in discussing or debating issues on a particular subject, like Shakespeare, the Cold War, or trout fishing. Not unlike the newsgroups of USENET, the communities created through BITNET provided social interactions without geographical boundaries. Like the ARPANET and the early Internet, BITNET was governed by specific standards and policies, including a policy that prohibited commercial communication across the network.

BITNET reached the height of its growth in 1989 when it connected roughly 500 organizations and comprised nearly 3,000 nodes. The growing pervasiveness of TCP/IP and the civilian Internet in the early 1990s eroded its membership; and in 1991 it was merged with a TCP/IP based network called CSNET, which is described next. A program written for BITNET called LISTSERV, however, survived the demise of BITNET by adapting itself to the growing and changing networked environment. LISTSERV coupled mailing lists with a file server (a computer configured to house, organize, and distribute files) and was engineered to help build network-based, virtual organizations. From the user's perspective, it relied on nothing more than email. But LISTSERV was able to host an open forum, moderate a discussion group, circulate an electronic newsletter, archive email correspondence, and even provide an electronic library.

## CSNET

Another network that came online in 1981, like BITNET, was the Computer Science Network (CSNET). CSNET was built by and for computer scientists who did not have access to the ARPANET; and it was funded by the National Science Foundation (NSF), an organization that would later be instrumental in taking over civilian operation of the ARPANET and in helping establish the commercial Internet. Two years later, in 1983, a gateway was installed to allow data to be transferred between the ARPANET and CSNET. The bulk of this data was email. In 1989, CSNET and BITNET merged to form the Corporation for Research and Education Networking (CREN), not to be confused with CERN, which is described below. Note: CREN was formally dissolved in January, 2003.

## CERN

During the 1970s and 1980s, a slightly different networking story was unfolding at the European Organization for Nuclear Research in Geneva, Switzerland, also known as the Conseil Européan pour la Recherche Nucléaire (CERN). (Not coincidentally, CERN was the birthplace of the World Wide Web.) The environment at CERN looked strikingly similar to that encountered by the early ARPANET team, except in miniature. It consisted of large, expensive mainframe computers located in a data center, isolated from each other, and used primarily to analyze large quantities of experimental data through a tape-driven batch processing approach. Since these data tapes were transferred by the physicists from their labs to the data center by bicycle, there were more than a few jokes about the 'network bandwidth' they enjoyed in being able to transfer so much data in such a short amount of time.[3]

Two CERN computer networks came into operation in the mid 1970s. One, called CERNET, was developed locally and provided data transfer from the labs to the mainframes in the data center. The other, called the TITN network, controlled one of the large particle accelerators. CERNET greatly resembled ARPA's Internet plans in that it relied on packet switching and focused on network robustness and reliability while requiring as little human intervention as possible. By 1978, Phase 1 of CERNET was complete, linking the main data center to the laboratory's remote site, the North Area. Two years later Phase 2 was complete, connecting over 50 computers across the two main sites at CERN.

Nearly all the networking hardware, protocols, and even operating system code that composed CERNET was developed at CERN itself. So when networking standards, such as TCP/IP and Ethernet, started to emerge and gain acceptance in the 1980s, engineers at CERN adapted their proprietary systems to internetwork with the growing number of Ethernet LANs that were springing up around CERN. This was their first experience with interconnecting dissimilar networks. Because of the international nature of the environment at CERN — with physicists coming from all over the world to work there — and the new hardware and software networking technologies coming into play in the 1980s,

CERN became a focal point in Europe for the development and integration of new networking technology.

CERN was also the principal link between the U.S. and Europe during the early years of the Internet; and, by 1990, it had become the largest Internet site in Europe. (An Internet site is a computer connected to the Internet that provides services and/or information accessible through one or more Internet navigation tools, such as FTP, RLOGIN, or a Web browser.) Many people at the time associated CERN so closely with networking that they considered the abbreviation to stand for: Center for European Research Networking. As will be discussed later in this book, CERN was an ideal breeding ground for the Web.

All of these networks, and others like them, succeeded in expanding the influence of computers and the power and potential of computer networks. As more people joined bulletin boards, exchanged email, participated in newsgroups, and started to access computers remotely, the networks grew and their influence widened. All that was wanting to bring some sense of unity to these seemingly isolated services was a large enough superstructure, and commercialization.

# The Transition of Control and Purpose

It took a U.S. agency named the National Science Foundation (NSF) to exert some control and direction over the loosely connected, seemingly free-forming Internet of the 1980s, and to perform some overdue consolidation. Established in 1950, the NSF was generally mandated to promote science and, in so doing, contribute to the prosperity, welfare, and security of the nation. Starting in 1977 and continuing through the 1980s, the NSF provided more and more funding and leadership for creating and maintaining the initial three-tiered Internet composed of universities, non-profit organizations, and government agencies. By 1985, the NSF was providing the bulk of the funding for the operation and continued development of the Internet.

The NSF began building its own network, the NSFNET, in the 1980s to link its supercomputer sites and to provide wider access to these large, expensive resources. The goals, planning, and development of the NSFNET were similar in scope and design to those of the ARPANET, with one important difference: the initial design of the NSFNET consisted of two distinct types of networks, rather than the single, homogenous network that constituted the ARPANET. Each of the university computer centers would be connected to a network covering a local, geographic region; while each of the regional networks would be connected over a central, high-speed backbone (i.e., a principal transmission line that operated at higher bandwidth and, therefore, could carry more data in less time). Because of this fundamental difference in design, the NSFNET more closely resembled the early Internet with its interconnection of dissimilar networks than the ARPANET.

The NSF understood the importance of regional networks. From the late 1960s and through the 1970s, the NSF provided funding for dozens of small, regional networks to allow universities to share computing resources and to help establish a more level playing field between well funded, larger universities and those with shallower pockets. With the creation of the NSFNET in 1986, the NSF increased its support for additional regional networks. By 1988, it had fostered the creation of seven new regional networks across the U.S., from San Francisco to New York to Texas to the Rocky Mountains. These networks, and others connected to the NSFNET backbone, together formed an extensive network that crisscrossed the U.S.[4]

The initial NSFNET backbone, however, exhibited two deficiencies that demanded swift remediation. The first deficiency involved the speed of the leased lines. They carried only 56 Kilobits (i.e., thousand bits) of data per second (56 Kbps), the same speed available on the ARPANET. Two upgrades to the lines over a period of roughly five years addressed this deficiency. By 1988, these lines were upgraded to T1s, which carried 1.544 Megabits (i.e., million bits) of data per second (1.544 Mbps), thereby increasing the transmission speed by a factor of 27 and enabling the lines to transmit something like 50 pages of text each second. By 1993, the backbone lines were upgraded yet again, this time to T3s, which carried 45 Mbps, increasing the transmission speed by a factor of 30 this time and enabling the lines to carry something like 1500 pages of text each second.

The second deficiency was that, unlike the ARPANET and the existing Internet, the NSFNET did not use TCP/IP. To remedy this problem and migrate the NSFNET's backbone to use TCP/IP, however, required the backbone to be taken offline for a considerable period of time. Arrangements were made that, while the NSFNET backbone was being upgraded, the NSFNET would use the ARPANET as its backbone. This connection, seemingly insignificant at the time, decisively and permanently changed the makeup of the Internet by making it available to most of the universities in the U.S. for the very first time. The Internet suddenly included non-profit regional network operators and a variety of new supercomputer facilities.[5] Here was the true beginning of a single, all-encompassing Internet.

The connection between the NSFNET and ARPANET's backbone also marked the beginning of the end for the ARPANET. The ARPANET was nearly twenty years old by the late 1980s and the new NSFNET backbone that was being built offered an ideal solution not only to transition the Internet to the latest technology but also to remove the military's involvement from what had become a largely civilian enterprise. This transition, which took place over 1988 and 1989, was facilitated by the fact that TCP/IP kept the network itself simple by having the host computers handle the more complicated functions. The vast majority of ARPANET and Internet users were never aware of the switch; and, in February of 1990, the ARPANET was shut down. All that separated the Internet of 1990 and today's Internet was completing the migration to privatization, that is, getting the NSF out of the NSFNET, and allowing for commercial traffic.

# The Arrival of Commerce

A workshop was held at Harvard University in March, 1990, sponsored jointly by the NSF and the U.S. Congress Office of Technology Assessment, to discuss commercialization of the Internet. During the workshop, commercialization was defined as "permitting commercial users and providers to access and use Internet facilities and services."[6] Privatization of the Internet, however, was considered a separate matter altogether. It was defined as "the elimination of the federal role in providing or subsidizing network services."[7] Both privatization and commercialization needed to occur before commerce could arrive on the Internet.

In 1983, the Department of Defense split the early Internet into the MILNET (for strictly military use) and the ARPANET (for continued scientific research). This was the first, albeit small, step towards commercialization of the Internet. Over the course of the next several years, control of the Internet passed from the Department of Defense to the civilian authority of the NSF; and traffic for the Internet migrated from the ARPANET's backbone onto the backbone provided by the NSFNET. Even after the Internet came under civilian control, however, commercial traffic remained prohibited. That's because the NSF was a government agency; and the NSF's Acceptable Use Policy (AUP) clearly stated that commercial traffic across its backbone was prohibited and that use was limited to educational, research, and non-profit activities. This policy, however, was considered very difficult to enforce; and it was clear to all involved that the ever-increasing number and types of activities and users on the Internet, and the increasingly vocal demands of corporations eager to offer services, meant that commercialization of the Internet was inevitable. What remained unanswered was how and when this change would occur.

The events that brought about privatization of the Internet, just as with many other significant events in the history of the Internet, were less the result of long-term planning than of the timely, but independent efforts of various individuals, groups, and organizations. Because the new technology of networking allowed people to create all sorts of new networks to meet their needs, the regional network operators that collectively formed the NSFNET, and other operators, also created independent commercial networks to meet the needs that the NSFNET could not satisfy. Over time, the operations of these network providers grew, some merged to form larger networks, and many established agreements to share and route traffic between their various networks. The result was an internetwork of commercial TCP/IP networks that paralleled the non-commercial Internet run by the NSF.

By the late 1980s, committees were formed to examine the use of the existing Internet and propose a process whereby the NSF could make the transition from an operation funded by the government to a commercial service.[8] These committees generally acknowledged that it was less costly and more efficient for the U.S. government to remain involved and continue to fund the Internet backbone. Plus, they warned that private enterprise would not be concerned with the network as a whole, but only with the more profitable areas and services. Nevertheless, they pushed for full privatization and wholesale commercial use of the Internet. Interestingly, although this proposed change of operation should have warranted a good deal of public debate and open discussion — after all, taxpayer money had been responsible for the creation of the ARPANET and the building and management of the early Internet — little or no public debate took place.

While privatization of the Internet was clearly the goal of the committee members, the first step in that direction focused on making changes to allow for commercial use. Some members noted that, regardless of the restrictions on commercial use, such use had been increasing 15-20% each month.[9] Subtle changes to the wording of the NSF's AUP in June, 1990 gave tacit approval to limited commercial use of the Internet, providing the NSF with a little more time to answer the growing demands for full commercial access. The solution for how to transition to privatization and, therefore, to full commercial use of the Internet — at least on paper — came in November, 1991, with the release of a new Project Development Plan from the NSF. The plan called for Internet service to be provided by independent Internet service providers (ISPs), as described in the following section. In short, each ISP would operate its own network and backbone, and gateways would be created to route traffic from one backbone to another, thereby replacing the need for the NSFNET backbone, any involvement from the NSF, and, in the process, eliminating entirely the commercial restrictions imposed by the NSF's AUP.[10]

## ISPs Step In

The subtle, but profound Internet policy changes in 1990 and 1991 led to the creation and widespread growth of companies called ISPs. The majority of ISPs were small and regional, providing local dial-up modem access for home or business computer users. Having an ISP account meant having access to the Internet, the same sort of access (albeit slower) enjoyed by a select group of researchers, government personnel, and computer-savvy university students. The access provided by the ISPs, like that provided earlier by bulletin board services, was primitive by today's standards. There were, for instance, no browsers and no graphical user interfaces. But you could send and receive email, browse bulletin boards, chat with people, play games, and participate in newsgroups, all without any geographical boundaries. Consequently, 1990 marked the opening of the floodgates for Internet access and use, even though relatively few people realized it at the time.

The same Internet policy changes that brought about the creation of ISPs also led to changes in the operation of the larger, commercial service providers that were then running regional commercial networks to satisfy the needs of local businesses. In 1991, three of these early, commercial service providers — PSINET, CERFNET, and ALTERNET — formed a non-profit organization called the Commercial Internet Exchange (CIX) that allowed them to interconnect their networks to form a type of network backbone. What they created was essentially a commercial version of the Internet, parallel to but separate from the government-run NSFNET. There was a membership fee and each member agreed to accept traffic free-of-charge from all other member networks. Aside from the fact that setting up some type of accounting system to determine how much to charge for traffic being passed from one network to another would have been a nightmare, customers would not have accepted service fees for network access that were similar in nature to telephone usage fees. Meanwhile, the CIX greatly increased the value of each of the member networks and, consequently, more and more networks in the U.S. and around the world joined, creating an ever-increasing commercial backbone.[11]

The NSF viewed the growing commercial networking industry as the fastest and simplest means to migrate the Internet to full, commercial use, allowing them to eliminate the government supported and managed NSFNET backbone and thereby arrive at a fully privatized Internet. Since the same companies providing the regional networks for the NSFNET were also behind these commercial networks, this greatly simplified the NSF's implementation of their 1991 plan to replace the NSFNET backbone. The transition was completed in 1994, when the ISPs and the regional commercial networks provided all Internet access through their own backbone and/or connection to part of the existing commercial backbone. Consequently, on April 30th, 1995, the NSFNET backbone was officially shut down.[12]

The shutdown of the NSFNET backbone left the U.S. government without its own research network, so it created the very-high-speed Backbone Network Service (vBNS). This network was restricted to government-controlled scientific research, but it was connected to the Internet through several gateways to facilitate the free exchange of information to which the researchers at government installations had become accustomed.[13]

## Early Internet Commerce

With the privatization of the Internet and the broad-based migration to TCP/IP successfully accomplished, the various, largely isolated groupings of network users started to interact and converge. Individuals who received their network access through online service providers like Compuserve, America Online, and Prodigy discovered the richer, wider resources of the Internet, either by switching to an ISP or by using a newly added Internet access feature offered by the online providers. Many of the most popular services provided by companies like America Online, such as chat rooms and individual or group instant messaging, started to appear on Internet sites. Similarly, the specialized services of UUCP, BITNET, and other networks, such as newsgroups and other types of moderated discussions forums, started to appear on Internet sites.

One of the first companies created exclusively to focus on the commercial Internet was CommerceNet, which was formed in 1994 with leaders from academia and the business world. Their mission statement reads:

> To transform the Internet into the world's largest and most efficient marketplace dramatically changing how the world transacts business.

Their early work helped convince credit card companies to accept a common security protocol for online transactions. They also helped establish, as well as enforce, guidelines relating to privacy issues for users of the Web. Their current projects are focused on developing standards that will increase the sharing and re-use of information across companies, and even across industries.

Also in 1994, a company called First Virtual put in place the first secure Internet payment process. Unlike the systems in use today, that employed by First Virtual did not use any data encryption or security-based protocols to ensure the safe transmission of financial information, such as credit card numbers, over the Internet. Instead, they used a decidedly low tech, but clever and efficient system that actually kept financial information secure by keeping it *off* the Internet.

Here's how it worked. You used the Internet to access First Virtual's site, and on that site you created a basic account called a VirtualPIN. Then, you called an 800 number to activate the account and input your credit card information. Transactions relied on email to take care of the necessary validation of information. When you made a purchase, you supplied the vendor with your VirtualPIN. The vendor then sent the transaction to First Virtual, which in turn sent you an email to confirm the purchase. Once you replied to confirm the purchase, your credit card was debited for the purchase. The First Virtual merchant was charged a flat cost of $0.29 per transaction, plus 2% of the transaction amount. The cost for an individual to use the system was $2 per year.

In 1994, Pizza Hut added an online order form to its Web page, marking the arrival of ordering food over the Internet. Also in 1994, the first flower shop began accepting orders over the Internet and online shopping malls started to appear.

The commercialization of the Internet that started in 1990 resulted in swift and permanent changes to the makeup of the Internet. Then, shortly afterwards, the Web arrived. As discussed briefly in the next section and at length in the second book in this series, "The Information Revolution," the creation and astounding proliferation of the World Wide Web unexpectedly and single-handedly transformed the Internet into something far more accessible, usable, and empowering. The Web also provided the perfect facility for introducing and proliferating all types of commercial ventures on the Internet.

## The Arrival of the Web

The arrival of the World Wide Web in 1991 helped considerably to widen and hasten the effects of commercialization on the Internet, a process that continues unabated today. But that's only a small part of the Web's story, and one that is far removed from why the Web's inventor, Tim Berners-Lee, set out to create his World Wide Web in the first place.

The Web was designed to be a network-based, document sharing and publication tool, something that would allow researchers and others to make their work more readily available over a private network, or across the Internet. It also included a means, through hyperlinks (electronic cross-references), to interconnect documents, thereby forming of web of interconnected references. Once it had been demonstrated, it took very little time for people to realize the inherent simplicity and power of the Web's design. One particular facility, for instance, proved instrumental to the Web's early and eager acceptance and, at the same time, provided a quick and easy solution to one of the most persistent complaints about the Internet — its lack of locatable resources.

The Web allowed programmers to bridge the gap between the new, interconnected world of networking and existing legacy systems (computers and applications designed with no notion of networking or remote access) that stored the vast majority of information. A company's phone book or database of documents, for example, that were only accessible from computers at one location before the Web, and often required customized software to first be installed on those computers, could, through the creation of a Web page and some relatively modest programming, be made accessible to anyone at any location who had a Web browser installed and Internet access. It didn't even matter which hardware platform or operating system was being used to access information made available through the Web. Before the Web, however, these computer system specifics often represented significant factors in determining which information systems someone could, and could not, use. Seen from the perspective of information access, the arrival of the Web was, in a word, liberating.

In 1993, less than two years after the Web's introduction, Mosaic, the first graphical Web browser to acquire a broad distribution, was created at the National Center for Supercomputing Applications (NCSA) located at the University of Illinois. Then, in 1994, the Netscape Corporation was founded, where many of the same programmers who wrote the code for Mosaic took on the task of outdoing their earlier work in order to produce Netscape's Navigator Web browser. The release of Navigator marked the beginning of the commercial browser wars. Over the next couple of years, Microsoft and other corporations quickly entered the fray, as browsers became standard operating system applications and later were integrated directly into some operating systems.

Browsers began as a friendly, easy-to-use window onto the Internet, but quickly evolved into a seemingly all-consuming application. By 1993, thanks largely to Mosaic, traffic on the World Wide Web was proliferating at an astounding annual growth rate of 341,634%. By 1995, the year Sun Microsystems announced Java and Compaq began shipping its personal computers with the Netscape browser, the Web surpassed the File Transfer Protocol (FTP) as the service responsible for the most Internet traffic. In a few short hectic years, the Web had apparently transformed the Internet; and yet, under the surface, the Internet itself, its basic services and design and its core architecture, had not changed at all.

The creation of the Web in turn sparked the creation of a whole new industry, one devoted to building applications and content for the Web. As these businesses became established and traffic over the Web began to increase exponentially, more and more existing businesses took notice and, through the Web, brought their business onto the Internet. Additionally, new types of Internet-only businesses emerged, starting what became the dot com boom (and subsequent bust, or bubble) of the late 1990s.

Amazon.com is a good example of just such a business: it relies entirely on its Internet storefront, presented through the Web, to bring products to consumers. Yahoo! is another Internet-only business. In contrast to Amazon.com, Yahoo! is an information service provider, generating revenue through pay services and advertising and providing a rich and diverse selection of information sources, such as news, weather, and stock information, and information management tools, such as a calendar, online bill paying, and email access. It, too, is totally dependent on the Web and its Web-based consumer interface.

Thanks in large part to the growing number of features packaged into browsers like Netscape Communicator and Microsoft's Internet Explorer, the majority of people today interact with the Internet primarily, if not exclusively, through one of these browsers. Browsers were designed to be easy to use, which contributed to their fast acceptance; and, perhaps as importantly, they were, for the most part, free for the taking. The Web itself, meanwhile, simplified the process of providing information sources and services on the Internet, adding greatly to the wealth of resources available over the Internet and expanding the application of those resources (e.g., not only could you access up-to-date data on a stock's price, you could also configure a tool to notify you when that stock reached a set price). Plus, the Web's hyperlink technology, interconnecting information between Internet sites and within an Internet site, contributed to making the Internet far easier to navigate and established a perfect complement to the Internet's architecture of interconnecting computers across any number or types of different networks.

In short, the Web and Web browsers were responsible for bringing a much larger, more diverse audience and a richer, more diverse selection of information content to the Internet. New people, resources, and services were arriving on the Internet at a rate far faster than anyone could have predicted.

## Standardization and the Battle for Network Control

Once it was clear that there was a future in networking computers, the quest for control of the burgeoning industry of networking products and services developed fast and furiously in the U.S. and Europe. From the point of view of many corporations and organizations, control of the industry became synonymous with control over networking standards and, more specifically, control of the process for creating those standards. International standards were nothing new in the 1970s. Consider the telephone, for example: if you knew how to use a telephone in one country, you could easily use one in just about any country. This was not coincidence; this was done by design and represented the result of conformance to international standards. The newness of computer networking in the 1970s, however, meant that networking standards, and organizations that served exclusively to create and support such standards, did not yet exist. It did not take long for this situation to change.

During the 1970s and 1980s, several very different groups, all interested in exerting control over which standards would be adopted and popularized, fought to protect their interests. These groups included: computer manufacturers, computer users, telecommunication carriers, and government agencies. Each group viewed the Internet from a distinctly different vantage point. Most computer users, for instance, were interested in perpetuating the type of grassroots involvement and influence that had already become one of the hallmarks of the Internet. For them, open, non-proprietary (i.e., not dominated or owned outright by any single company) standards and a standards process that welcomed individual participation were key to keeping the Internet free to grow and develop as its community of users wanted. Computer manufacturers, on the other hand, needed clear and well-defined standards in order to build and promote Internet-ready computers. Multiple standards would result in a fractured marketplace with increased costs in order to meet the different needs of different customers. A single, proprietary standard would force each manufacturer to submit to the mercy of the one corporation that owned the standard. Computer manufacturers, therefore, wanted a voice in the process of creating the standards, so that they would know where the industry as a whole was heading and be ready to market compatible products at the earliest possible moment.

One of the earliest and largest conflicts in the Internet's struggle for standardization pitted the computer manufacturers (the sellers) against the telecommunication carriers (the buyers) over who would be the controlling force behind the growing networking products industry. Which group, for instance, would own the specifications for the network cards that each Internet-ready computer needed in order to connect to the network and exchange information? The telecommunication carriers provided the service and equipment that carried all the data on the Internet. But the computer manufacturers created the machines that embodied the network and provided all the information and services on the Internet. Directly connected to this struggle was the simple and straightforward question of where responsibility for the quality of service on the network resided: was it with the network (meaning with the phone companies) or with the host computers (meaning with the computer manufacturers)?

Many saw this conflict as one that would determine who would eventually control the Internet itself. That's because the conflict centered on which standards would constitute the core network architecture of the Internet and, accordingly, who would control those standards. Would it be the existing, de facto standard of TCP/IP, which originated with ARPA's Internet project and to which no one organization could claim ownership or advantage? Or would it be a competing standard, a set of three protocols designated Recommendation X.25, that the telecommunications industry succeeded in getting created and adopted.

X.25 was developed by representatives of telecommunication carriers from Canada, France, and Great Britain, as part of a working group in the international standards organization called the Consultative Committee on International Telegraphy and Telephony (CCITT). In September, 1976, X.25 was approved, and the telecommunication carriers immediately started to incorporate the new protocols into the data networks they were building. From a strictly technical standpoint, the telecommunication carriers could easily have adopted TCP/IP for their networks, saving themselves the considerable time, energy, and money that went into the development of X.25. Moreover, U.S. representatives to the CCITT suggested that they do so. But the suggestion was quickly rejected. This was more a political and financial battle than a fight over the specific strengths and weaknesses of the individual protocols.

The X.25 and TCP/IP protocols were both written to accomplish the same task: to provide the low level infrastructure to network computers together. But whereas X.25 was a *connection-oriented* network in which the the packet switches of the subnet were, much like the IMPs of the ARPANET, responsible for maintaining order and control over the packets of data, TCP/IP was a *connectionless* network in which the host computers provided the control and corrected errors, making few assumptions or demands about the reliability of the network itself. This difference is easily understood, given that the telecommunication carriers were in the business of ensuring reliability across their networks, but were indifferent to what was happening at the sending and receiving points, while the computer manufacturers (and the Internet engineers), on the other hand, wanted to accommodate as many different types of networks and network conditions as possible, including those that were inherently unreliable, like packet radio, and those used by the military, which might undergo disruptions in service.

In very general terms, X.25 was designed with expectations that were perfectly reasonable for monopolistic telecommunication carriers: the network was reliable, control was an integral part of the network, as few demands as possible were made on the end users and their equipment, and networks would be largely homogenous (all using X.25, for instance). The early Internet community, however, faced very different circumstances and designed TCP/IP to meet those needs: network reliability could not be assumed, end-to-end control was essential since only control at the hosts could ensure retransmission of lost data, many different network types were already emerging, and being able to interconnect dissimilar networks was key to the future growth of the Internet.

The telecommunication carriers succeeded in creating an alternative to TCP/IP with X.25, another distinctly different type of network, complete with its own distinct set of strengths and weaknesses. But X.25 failed to disturb the widespread acceptance and use of TCP/IP, which was readily adopted by most computer makers and manufacturers of networking components in the 1970s and became a de facto commercial standard in the 1980s. The grassroots beginnings of TCP/IP and its inclusive architecture of facilitating the interconnection of any and all types of networks, including X.25 networks, best represented what most Internet users perceived as the Internet's greatest strength: its innate ability to thrive and grow without any single controlling authority and without any single set of rules. The creation of X.25 did have at least one lasting effect on TCP/IP. It made clear the need for a standards' process to keep TCP/IP open and to enable it to evolve as networking conditions changed. More information on the standardization of TCP/IP will be presented later in this book.

## Standardization and the International Arena

On a different front, another international standards body, the International Organization for Standardization (ISO), tried to address the issue of incompatible computer network products. Each manufacturer was producing its own proprietary products that functioned only with its own computers. ISO members from the U.S, Great Britain, France, Canada, and Japan decided that an open systems approach was needed, and it should consist of public, technical specifications in order to encourage computer manufacturers to create compatible products. Changes would be issued by a public standards organization; and design would focus on generic components, rather than on one company's proprietary model. This approach gave consumers considerable influence over the network products that would be built. ISO called their project the Open Systems Interconnection (OSI).

What ISO created amounted more to "a standard for creating networking standards" than to technical specifications that a company could read, interpret, and use to build products.[14] The OSI model arranged the various network functions into seven layers of protocols. The first layer, the physical layer, was the most concrete and dealt with the transmission of electrical signals. The link layer was next; and it covered translating the electron flow from the physical medium into the bits of data, and determining when to send and receive messages. Ethernet is an example of one type of link layer protocol; and it later became an official OSI standard for the link layer. Each layer in the OSI model became progressively more abstract, with the top layer handling the applications, such as FTP and email.

The greatest achievement of OSI was that it provided a framework and a vocabulary that everyone could use when discussing and engineering networking products. Many existing products were revised to comply with the OSI model; and even companies like IBM, which relied on the proprietary nature of its products to retain market share, had to offer products that conformed, in order to meet consumer pressure. In the mid 1980s, both TCP and IP would separately become international ISO standards as TP4 and ISO-IP, respectively.

For a variety of reasons, however, the OSI model never had the kind of impact its creators intended. Basic compatibility between networking products from different manufacturers remained more the exception than the rule; and proprietary network systems still thrived. From the 1980s on, the number of manufacturers producing networking products has continually increased, as has the the diversity of these products, the protocols that accompany them, and the types of networks they are used to construct. In the end, and very like the telecommunication carriers development of X.25 networks, the international attempts to establish standards served to broaden the number and types of networks rather than to establish a more homogenous network environment. The only thing capable of providing some measure of compatibility between all of these different networking protocols and products, and interconnecting so many different types of networks, was, and is, the Internet itself.

# The Bare Essentials of Management

When the U.S. government transferred all Internet backbone traffic from the government-controlled and government-financed NSFNET to the private, commercial sector, the questions that were asked then, and are still asked today, focused on the Internet's management: who is in charge, who is making the rules, who has control and authority?  The general and most accurate answer, then and now, remains: "rough consensus and running code."[15]

This deceptively simple answer strikes right to the heart of how the Internet is governed and who really is running it.  That's because the operation of the Internet is managed and controlled by the code that defines and implements the interoperability of the network: its running code.  This code can be found in every computer and network device that functions on the Internet; and TCP/IP constitutes a large part of this running code.  As to the rough consensus, that relates to the participation of individuals and industry representatives from around the world who make the time to debate and vote on the various technology issues related to the evolution of the Internet.  It also relates to what ordinary users of the Internet demonstrate, through the sheer power of their numbers, to be the services, functions, and features of the Internet that best fulfill their wants and needs.  There was, for instance, no discussion, planning, or voting related to the introduction of email.  Email became a service on the early Internet, quickly found an audience, and its widespread acceptance and use changed the evolution of the Internet.

Management of the Internet also comes in two other, more specific forms.  One form consists of a collection of advisory and standards bodies that function to discuss and plan for the Internet's future.  These organizations will be described in chapter 8.  The other form consists of the system and the organization charged with managing the day-to-day operation of the Internet, both of which are described below.

## Domain Names: Identifying and Locating Internet Hosts

The day-to-day operation of the Internet relies in large part on the Domain Name System (DNS), which was introduced in the previous chapter.  Just like a telephone number or a postal address, a computer's address on the Internet must be unique; and just like the systems that locate and ring the telephone identified by its number or deliver a letter to the address on its envelope, the Internet requires a system to route and deliver data to the computer identified by its address.  DNS serves both of these functions, identifying and locating the millions of computers that connect to the Internet every day.

A computer's identification consists of a unique, easy-to-use name (useful for us humans) called its domain name (e.g., lobsters.downeastmaine.com) and a corresponding, unique number (necessary for the computers) called its Internet Protocol (IP) address (e.g., 115.99.112.28). DNS is responsible for storing all Internet domain names and IP addresses in its own, specially designed distributed database (think of a large ledger or file cabinet, with individual pages or file folders distributed throughout the Internet). DNS is also responsible for the system and policies that allow this information to be retrieved and updated as needed.

When you send an email message to the address help@siamesecats.com, for instance, DNS is employed as part of the process of delivering your email. DNS translates the domain name, siamesecats.com, into its corresponding IP address by sending a request to a specialized computer, called a name server, which looks up the IP address for the given domain name in tables it has stored locally. If it can't find the domain name, it forwards the request to another name server and awaits its reply. That server repeats the process, first checking its locally stored information and, if it fails to locate the domain name's IP address, then forwarding the request to yet another server. This process is repeated until the IP address for the domain name is found, and it is returned to the computer that made the original request. Only after the IP address has been retrieved can data be routed across the Internet to the specified destination and the email be delivered.

Each and every minute of every day computers come and go on the Internet. Some have permanent IP addresses, which are known as static IP addresses. Others have an IP address that can change every time they connect to the Internet, which are known as dynamic IP addresses. For the Internet to function, for data to be sent across the Internet and received at its intended destination, all of these computer names and addresses must be stored somewhere, and this information must be continuously updated. DNS, therefore, maintains a level of fundamental and necessary order across a highly dynamic and distributed network. Without DNS, or some other equally efficient, hierarchical system to identify and locate the millions of computer systems and networks that compose the Internet, there could be no Internet.

DNS, like many other fundamental components of today's Internet, didn't just miraculously spring into operation. It grew out of an earlier methodology used to impose the same general type of order, but that ultimately proved itself inadequate to the task as the Internet grew and evolved. Throughout the 1970s and well into the 1980s, it was one of the tasks of the Network Information Center (NIC) located at the Stanford Research Institute in California to approve and allocate computer names and addresses and to maintain and distribute a single, master table of computer host information. When the Internet comprised a few dozen to a few hundred computer systems, such a system was relatively fast and effective; and it did the job of maintaining order. As the network grew, however, it quickly became apparent that this system was grossly inadequate:

- It would not scale as the network continued to grow.

- As the host information file grew, it used more and more network resources, and it took more and more time, to distribute the file across the network.

- The NIC could not update the information as quickly as was needed, which led to local administrators updating their own files and throwing the system out of sync.

Solving these problems led to the initial design and creation of DNS in the early 1980s. To make it possible for the system to scale as the network expanded, DNS divided the network into hierarchical subsets. This involved dividing the network structure into seven principal, high-level domains, distinguished by three-letter domain name suffixes:

- Commercial (.com)

- Educational (.edu)

- Non-profit organization (.org)

- Government (.gov)

- Network (.net)

- Military (.mil)

- International (.int)

The domain for international networks, incidentally, reveals both the U.S.-centric thinking, or bias, of the engineers responsible for this dividing of the network and the existing U.S. government domination and ownership of the Internet. But this deficiency was soon corrected, as the naming conventions were expanded to include country-specific domains (e.g., .uk for United Kingdom and .fr for France) in order to accommodate the international growth of the Internet and better distribute control. Each of these high-level, or *root*, domains could then be divided into smaller and smaller subdomains, as necessary or desired. For instance, a large company like IBM has the principal domain, ibm.com, as well as many subordinate domains, like ones based on location, such as austin.ibm.com.

Just as the domain names themselves are hierarchical, proceeding from the most general (e.g., .com) to the increasingly specific (e.g., austin), so is the system of computers that holds the data associating a computer name with its IP address. Special computers called *root servers* are distributed across the network and form the highest level. The root servers contain information with the names and addresses of other specialized, DNS computers, called *name servers*. The name servers contain name and address information about other name servers, as well as name and address information about host computers connected to their local network. This hierarchy continues downward to name servers that contain the same type of information for individual companies, organizations, government institutions, ISPs, and so on.

This hierarchical and distributed organization of host name and address information is what allows local network operators to maintain control over their particular segment of the network. It is also what eliminates the need for the master database and single point of control that characterized the earlier system. As described above, each name server contacts the next name server in this hierarchy in the search for a particular host's IP address. This process repeats until a name server is located with the specific information, or until the local name server responsible for that host's domain has been contacted and the IP address is returned.

The system is so efficient that most lookups complete in the merest fraction of a second.

Unfortunately, despite the speed and efficiency of its operation and the generally distributed nature of its storage of information, DNS, like other hierarchies, requires some authoritative control at the very top. There is, accordingly, a single organization that controls the root servers at the top of the DNS hierarchy and that also controls the assignment of domain names. Without such an authoritative body there would be no way to ensure that there is always a one-to-one match between a computer name and its IP address regardless of how or where you are connected to the Internet. This organization is called ICANN.

## ICANN: Controlling the Allocation of Internet Host Names

The Internet Corporation for Assigned Names and Numbers (ICANN) was formed in 1998 as a non-profit, Internet-community-based organization. Its charter is to manage the root servers as a matter of public trust for the international community of Internet users. Its primary functions are to control the fair allocation of domain names and IP addresses, and to manage the root DNS operation and the root servers, functions that had previously been performed under a contract from the U.S. government by the Internet Assigned Numbers Authority (IANA) and other organizations. ICANN, like many other Internet-related organizations, is composed of an assortment of advisory committees and task forces that try to find common ground among informed users of the Internet and to derive policies based on perceived best interests.

In addition to keeping DNS running, ICANN accredits the organizations that operate as *registrars*, where individuals and organizations go to select a domain name for use on the Internet. ICANN is also the place where disputes over domain names — such as when someone has registered a domain name that is the trademark of an existing corporation — are resolved.

Even though ICANN has only been in operation for a few years, there are already controversies over the organization's control, especially outside the U.S. A current problem is the creation and use of root servers that are not operated by ICANN, a practice that could lead to name and IP address conflicts. As the Internet continues to grow, so will issues like this. But without some form of centralized control, like that provided by DNS and ICANN, the Internet will not be able to sustain itself as a truly global communication structure.

## The Internet Worm of 1988

Today, even if you don't own a computer or use the Internet, it's likely you've heard of something called a computer virus. Broadcast news agencies frequently report on some new malicious program, typically a new type of computer virus, making its way across the Internet. More often than not, such viruses are transmitted via email, although viruses can spread their infection through other means, such as shared computer programs and floppy disks. Viruses transmitted by email quickly infect thousands to hundreds of thousands of computers, and more. Some people end up losing current or recently completed work, while others find that program or operating system files on their computer have been corrupted or destroyed. These viruses can even cause computers to crash and no longer function. Each new virus can easily cost millions of dollars in terms of lost work, cleanup time, and damaged equipment. Networking technology and, more specifically, the Internet itself, makes such attacks not only possible but relatively easy (at least to attempt), and potentially devastating in terms of their consequences.

While security was an early and serious concern of the Internet's engineers, it was the Internet worm of 1988 that brought widespread attention to the issue. A worm, unlike a virus, is a self-contained program that makes its way from machine to machine across the network, using one computer's resources to find and attack others. Its intention is not typically to destroy files or cause permanent damage to computer systems. Instead, its purpose is to travel fast and far, and to reproduce. A computer virus, on the other hand, is not self-propagating, but it is designed to be destructive. A virus is typically a fragment of a program hidden inside another, larger program; and when you execute that larger program, by, for example, starting up Microsoft Word to view a document, the virus is unleashed.

Some people think of November 3, 1988 as Black Thursday. Computer system administrators across the U.S. came to work and, if they could log onto their systems, found their networked computers struggling under an unusually heavy load of activity. Overnight, a worm had invaded their systems from across the Internet. It was replicating itself faster than it could be destroyed, and migrating from system to system.

What is now known about the worm's attack is that Robert Morris, Jr., a graduate student studying computer science at Cornell, created the self-replicating, self-propagating program and released it onto the Internet from a computer at MIT on November 2, 1988. Apparently, there was a bug in his program and it started infecting machines and replicating itself at a much faster rate than anticipated, which caused the infected machines to crash, or to grind to a halt and become useless. Within a few hours the worm had infected thousands of computers across the U.S. With a friend from Harvard, Morris tried to send an anonymous message out over the Internet explaining how to kill the worm and prevent further contamination. Ironically, his worm had effected so much damage throughout the Internet that his message arrived only after the situation was under control.

Morris's worm exploited security lapses in several protocols and programs residing on computers that ran a version of the Unix operating system from Sun Microsystems called SunOS, and other Unix machines running 4.2 or 4.3 BSD (a variant of Unix called the Berkeley Software Distribution developed at the Berkeley campus of the University of California). These workstation computers (minicomputers more powerful than personal computers) were among the most popular networked computers at the time, so the worm affected many universities, military installations, and research facilities. In short, the worm was designed to cycle through each area left unprotected by the security lapses until it succeeded in connecting to other machines across the network, bypassing standard authentication, copying itself, and then continuing onward. It included program code that helped to keep it undetected by system administrators and that made it difficult to kill. It was estimated that the cost of destroying and cleaning up after the worm at each installation ranged from $200 to more than $53,000, depending on how many computers had been infected and how seriously each computer's system files had been compromised.

It took several days before some semblance of normalcy returned to the Internet. Many individuals and teams of programmers spent countless hours trying to debug the worm and determine what was needed to shut it down and prevent it from happening again. Morris was eventually found guilty of launching the attack. More specifically, he was convicted of violating the computer Fraud and Abuse Act (Title 18). He was sentenced to three years of probation, four hundred hours of community service, and fined $10,050 and the costs for his supervision.

There's little doubt that the Internet Worm of 1988 changed the Internet. The story was on the front page of the New York Times and other newspapers around the U.S. for days after the event; and it was discussed on television and radio programs, too. This was for many their first introduction to the Internet. Commercial use of the Internet was still a few years away and the Web was still a year away from even being proposed. For those who regularly used the Internet at the time, particularly those who relied on it for their work or whose work it was to maintain the computers that composed the Internet, the event was a loud wake-up call.

Imagine for a moment if the Internet were today made unusable for a week, or even a couple of days. The Internet and its infrastructure is considered by many individuals, businesses, and organizations to be a fundamental, even essential service. Already we depend on it every day, some for recreation, some for essential communication and security, some for a livelihood; and this dependence is fast becoming little different from our dependence on telephone and television service or on an uninterrupted supply of electricity or gasoline.

# Packet Switching: Lifeblood of the Internet

# 4

## The Digital World of Computer Communication

Before the Internet could evolve, before the ARPANET could be built, an entirely new communication system was needed, one that would make possible the reliable and efficient transmission of data in a network composed of computers. As explained in chapter 2, the only large-scale communication system that had been built when computer networking was first being explored in the 1960s was one that had been designed for the transmission of human communication. This was the telephone network, and it relied on a technology called circuit switching to accommodate the specific needs of person-to-person communication. Since the communication needs of humans and computers diverge so greatly, there was no way to adapt circuit switching to the needs of a computer network. Therefore, the solution to the problem of how to provide reliable and efficient data transmission in a computer network was found in the invention of a new network technology called packet switching, and the creation of packet-switched networks.

One element that distinguishes computer communication from human communication focuses on the fact that computers inhabit a strictly digital world. Computers communicate, operate, and represent information by means of a digital language, a binary or two-valued mathematical language composed entirely of sequences of zeros (0) and ones (1). Everything we do on the computer is at some point encoded and stored in this digital language. Whether it's simple text or numbers, graphics, an audio or video composition, or anything else you can think of, all computer data gets represented as some sequence, or series of sequences, of zeros and ones. Conversely, every calculation and operation the computer performs, every piece of data it stores, must be translated into, or represented in, a form that we can understand. The result of this transformation may be English or German text, or the image of a sleeping cat.

Consider something as basic as the letters of an alphabet. First, how do you represent these letters in a binary format of zeros and ones? Second, how do you know that the representation of this alphabet in your computer is the same as it is in other people's computers? After all, such decisions are abstractions, like an alphabet itself, agreed upon through some arbitrary process to facilitate communication. Some group or organization must first determine an approach or methodology that suits their needs, and declare, for instance, that an *A* will be represented by the following unique sequence of seven binary digits, *1000001*. Other groups may establish their own digital representation, which means in order to share data they must create a tool to translate data from how others have represented, or encoded, their information into the representation they have chosen. Or they accept one particular abstraction and thereby enable computers from the respective groups to more easily share data; and if this acceptance becomes sufficiently widespread, it may evolve into an industry standard.

The following table shows a representation of the uppercase letters of the English alphabet that is part of one such agreed upon standard, the American Standard Code for Information Interchange (ASCII):

| ASCII Alphabet | | | |
|---|---|---|---|
| A | 1000001 | N | 1001110 |
| B | 1000010 | O | 1001111 |
| C | 1000011 | P | 1010000 |
| D | 1000100 | Q | 1010001 |
| E | 1000101 | R | 1010010 |
| F | 1000110 | S | 1010011 |
| G | 1000111 | T | 1010100 |
| H | 1001000 | U | 1010101 |
| I | 1001001 | V | 1010110 |
| J | 1001010 | W | 1010111 |
| K | 1001011 | X | 1011000 |
| L | 1001100 | Y | 1011001 |
| M | 1001101 | Z | 1011010 |

ASCII was the first universal computer standard. It solved one of the earliest and most fundamental problems faced by computer engineers interested in sharing data among different computers, enabling the transfer of data in a common format between any type or make of computer that conformed to the ASCII standard. ASCII consists of 128 unique 7-bit sequences called strings; each string stands for a letter (uppercase and lowercase letters are represented separately), an Arabic numeral, a punctuation mark or symbol, or a function, like a carriage return. The word *bit* is a contraction of *binary digit*; each bit has a value of either 0 or 1. A sequence of 7 bits was required to accommodate all the elements that had to be represented. Other, larger standards came later. For instance, a relatively new language standard called Unicode is enormous, because it attempts to accommodate the needs of all alphabets and languages and therefore requires far longer binary sequences than ASCII's economical 7-bit strings. It wouldn't make much sense to network computers if the data being transmitted was encoded in a format unique to each computer system. So a standard such as ASCII was critical to getting the process started, because it provided a basic, but essential framework for the representation of information within and among computers.

Since computers encode information and perform operations digitally, it makes good sense that they should also communicate digitally. When computers communicate today, their digital messages are arranged into a discrete digital signal, or digital waveform. This signal is electrical and represents the zeros and ones as on-off electrical impulses. The signal is then transmitted from whatever transmission medium is on the sending computer, for instance an Ethernet card, onto the connected network. The computer on the receiving end takes the data through the same process, but in the reverse order.

When networking was first being explored, however, there was no such thing as a digital communications network. Why would there have been? Computers were isolated devices, and nothing else communicated in a digital format (certainly not people). The circuit-switched telephone communication system was and is analog, designed specifically to carry voice communication. It transmits waveforms of electrical sound impulses that are analogous to the frequency and pitch of our speech patterns. This is why, for instance, you need a modem when you use a phone line to connect to the Internet. The word *modem* is a contraction of *modulator-demodulator*. The purpose of a modem is to convert the stream of digital information coming from your computer into electrical sound impulses so that it can be transmitted across the analog telephone system. A modem at your ISP then reverses the process, converting the electrical sound impulses back into a stream of digital information that can be transmitted across the packet-switched, digital Internet. These sound impulses can often be heard soon after your computer's modem dials your ISP, as the two modems proceed to *screech* at each other in alternating pitches in their attempt to communicate and establish a connection.

This fundamental incompatibility between the digital communication of computers and the analog communication network of the telephone system meant that a new, fully digital communication network would need to be built. The engineering and construction of such a network, however, was just one of several significant obstacles that had to be overcome in networking the first computers, as will be explained below.

# Inventing a New Communication Network

In the late 1950s and early 1960s, when networking computers was first being explored, the only existing communication or network model was the telephone system. Think of an extensive railroad system with thousands upon thousands of track switches that could be configured into a virtually limitless number of different routes. This was the telephone system, with circuit switches performing the same function as the track switches. Each telephone call resulted in the creation of a unique and dedicated route between the caller's telephone and the telephone of the person the caller was trying to reach. The configuration of all the intervening circuits that created the route remained in effect for the duration of the call, and the direct line established by these circuits — their effect was no different than a single wire connecting both telephones — was reserved, or dedicated, for the exclusive use of that single conversation. This system was designed specifically for how people communicated: it reserved the completed circuit for the sole use of the two parties, and remained open, or on, all the time no matter how much or how little was being communicated.

When the telephone was invented in the late 19th century, circuit switching was done manually by operators. They used patch panels, which consisted of boards each containing some number of connected telephone lines, along with cords that they would plug into the panels to electrically connect one line with another, thereby creating a circuit. The operator local to the caller had to determine if the person being called was also local, in which case she would complete the circuit herself by connecting the two lines right there in the panel in front of her. (Nearly all operators were female.) Otherwise, she would connect to the next operator along the line, and this process would be repeated until the operator local to the person being called was reached and the entire circuit was completed. Over time, human switch operators were replaced by mechanical switches, and, eventually, these mechanical switches were, in turn, replaced by computerized switches. Nevertheless, the overall circuit-connection process and its result are essentially the same today as they were more than one hundred years ago.

The first data-communication networks used by computers in the early 1960s were composed of dedicated, analog circuits. They relied on computer modems and the telephone network to interconnect one computer directly with another, no different than how a personal computer today connects to an ISP through its modem and a telephone line. Once the data communication circuit was established between the two computers, it remained on continuously regardless of how little or how infrequently data was being transmitted. While this type of dedicated connection was necessary for voice communication and could be adapted to handle the transmission of data, this initial approach to networking was costly and inefficient. Moreover, it forced computers to translate their data through the modems from its native, digital format into analog sound impulses, and later back again into its original digital composition, in order to use the network.

A communication network for computers needed to operate in a fundamentally different manner. It needed to work asynchronously, that is, it needed to transmit data without being concerned about when that data would be received. Also, since computers, unlike people, communicate in brief, intense bursts using small, discrete packages of data, it needed to transmit data in small packets, and only when needed. The network, like the computers themselves, also needed to operate as a fully digital system. Moreover, a radically new network architecture was required to meet the needs of the military. The military was funding the computer networking projects and it demanded built-in redundancies and decentralization so that its communication network could withstand a nuclear attack. Meeting these requirements, therefore, called for something entirely new. This turned out to be packet switching.

Two engineers were effectively inventing packet switching at roughly the same time in the 1960s. Paul Baran, working at the Rand Corporation in the U.S., approached packet switching from a military perspective. He wrote a series of reports that were de-classified in 1964 and published under the title, "On Distributed Communications." In these reports, in which he describes a communications system designed to survive a nuclear attack, he presents detailed descriptions of several of the key elements in a packet-switched network. At about the same time, Donald Davies, working at the National Physics Laboratory (NPL) in the United Kingdom, approached packet switching from the perspective of computer networking. His interests, like those of the ARPANET engineers, derived from a desire to implement a time-sharing computer model. For him, packet switching offered a way to better distribute the high costs of expensive communication equipment while using this equipment more efficiently. Taken together, the work of Baran and Davies provided the ARPA engineers with the necessary models and theories to build the first packet-switched network.

The following sections briefly describe Baran's focus on the military application of packet switching and Davies' focus on the application of the time-sharing model to packet switching. The final section presents an overview of the evolution of packet switching, from its beginnings with the ARPANET through the implementation changes that helped bring about the creation of the Internet.

## Survivable Communications and Packet Switching

Baran focused on the need for the military's command-and-control system to remain operational even after a large percentage of the underlying infrastructure (i.e., the switches and wiring) had been destroyed. Baran saw the need for a system that — unlike the phone network, which was hierarchical in nature and could be disrupted by disabling specific switches - was distributed, with multiple and highly redundant switching nodes that could automatically take over circuits if and when other switches failed. He captured the different types of network node configurations in the following figure:[1]

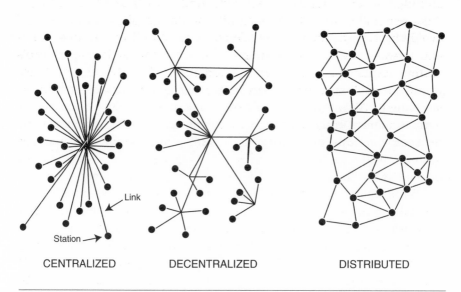

CENTRALIZED          DECENTRALIZED          DISTRIBUTED

**Figure 1. Network Node Configurations**

This figure powerfully illustrates his argument about the inherent vulnerabilities of any type of centralized, or even decentralized, network. It also clearly shows that the matrix of a distributed system would be the most capable of remaining functional after an attack, given its greatly increased capacity to reroute traffic to any one destination.

For Baran, a survivable communications network had to meet the following basic conditions:

- The network must remain functional even if a large number and/or several central nodes were disabled.

- Switching nodes must be placed far from population centers.

- The network must maintain excess capacity, in order to adapt better when parts of the network became damaged.

- The system must be able to handle encrypted messages.

- There must be a message priority system, in order to distinguish more urgent messages from routine messages.[2]

To satisfy these requirements, Baran proposed a system that included several hundred switching nodes with as many as eight links per node. In order to simplify the design of the switches and avoid the type of message bottlenecks exhibited by existing store-and-forward switching systems used by the military, Baran's system required a high transmission rate. This way the switches would need less storage, and message management would be simplified. Also, rather than broadcast messages (i.e., have every node send out each message to every other node to achieve the necessary redundancy), as others were proposing at the time, his system would intelligently route messages from one node to the next.

While Baran was working on his proposal, AT&T was building a specialized analog telephone network called the AUTOmatic VOice Network (AUTOVON) for the Air Force. The switches in the AUTOVON were arranged in a grid system, similar in structure to the matrix design of Baran's distributed system. But AT&T designed a single operations center where traffic was monitored by individuals. Furthermore, routes were changed manually, which left the system vulnerable to attack. In contrast, Baran's proposal included no operations center and specified that the switches themselves should determine the routes and adjust automatically to changes in the network. Baran also specified that data transmission should be digital, not analog as it was in the AUTOVON. Digital transmissions could be regenerated without compromising the data, whereas analog signals degraded as they passed from switch to switch. At the time, the digital world of computers was foreign to most people, including the telecommunication engineers at AT&T. Ultimately, it would take networking, the Internet, and finally the Web to bring about a complete digital revolution.

To implement the high transmission rate that he needed and still maintain reliability, Baran proposed that all message blocks (chunks of data) be one size, specifically 1024 bits. A short message could easily be contained in a single message block, while longer messages would be divided into a series of blocks. Each block would contain address and control information in addition to the data being sent; and each would follow its own independent route to the destination, where the data would be reassembled, if necessary.

Baran's inspiration for his concept that messages could discover their own way through the network to their destination came from mathematician and computer scientist Claude Shannon. Shannon's theoretical work was central to the digital computer revolution, but it was his creation of a machine for an existence proof that caught Baran's attention. With this machine, Shannon trained a mechanical mouse to find its own way through a maze, just as the packets in Baran's messages were to find their own route through the network.

Baran's distributed communication system was never built. AT&T was approached by the Air Force to build an experimental version of the system; but engineers at AT&T did not believe the system would work. When the project was moved by the Pentagon to the Defense Communications Agency (DCA), the agency's lack of experience with digital technology and computers resulted in the same disbelief. Realizing that the time was not right for trying to implement his system, Baran canceled the project himself rather than see it fail from lackluster support.

## Time-Sharing and Packet Switching

In 1966, Donald Davies coined the term packet switching, even though by then Paul Baran and others had been debating the concepts of data packets for some time. Davies' principal interest was to improve the economies and transmission rates of interactive, time-sharing computing. At the time, access to remote computers was over circuit-switched telephone lines. This meant that each open circuit was monopolized by a single computer-to-computer connection; and a significant amount of time and expense was wasted while the computers waited on input from the operator or on output from a running program. Davies realized that, for time-sharing computing to work, the methodology that allowed data to be transmitted between computers had to be built upon the same time-sharing concepts that allowed the resources of a single computer to be shared among different users, and for different operations, simultaneously. In other words, packet switching was to the network what time-sharing was to the individual computer: a means of sharing resources by taking into account the efficiencies of how computers operated, whether those efficiencies were to be found in how computers communicated in short bursts of data or in their capacity to switch so quickly between operations that they could multitask for multiple users.

In 1965, Davies wrote a proposal outlining a packet-switching network for the United Kingdom. The network would connect major cities over high-capacity telephone lines; the majority of its nodes would have multiple connections, and it would have a dynamic routing system that adapted to changes in the network, not unlike Baran's. The specifications for the network included features that would not be realized until the 1980s, some twenty years later: trunk lines with data transmission speeds from 100 Kbps to 1.5 Mbps (i.e., a T1 line), a message size of 128 bytes, and switches capable of processing 10,000 messages per second.

Unlike Baran's packet-switched specifications, Davies' system was designed for business and personal use, not for the military, and was meant to provide services like remote data processing and sales-related transaction processing; it even included online betting. The General Post Office, which controlled telecommunication in the U.K., endorsed Davies' proposal. Funding, however, was insufficient for Davies to build the scaled-down system he designed for demonstration purposes at the National Physics Laboratory. A single-node network was all that was created. It functioned as a test bed for networking ideas and became one of the first local area networks.

# Packet Switching: From ARPANET to Internet

It took the creation of the ARPANET to put the hard work of Baran and Davies to the test. The original four-node ARPANET established in 1969 was the first practical application of a multi-node, packet-switched data network. This early, experimental network bore little resemblance to Baran's highly redundant, distributed communications network. The dissimilarity was due not to its design, however, but to its limited size and scope. Even though the early ARPANET was very small, it was unequivocal proof that the fundamental efficiencies and design goals of packet switching detailed by Baran and Davies could be realized.

Success, however, was not immediate. Moreover, as is true of much of the early work in networking design and engineering, the implementations of packet switching evolved over time, as theory was applied to a working network and that network started to grow.

The story of the ARPANET and packet switching begins in 1967. In that year, at the ACM Symposium on Operating System Principles in Tennessee, Davies presented a paper on packet switching. At the same conference, Lawrence Roberts from ARPA presented a paper on computer network design. Also attending the conference was Leonard Kleinrock. Kleinrock, a graduate student at MIT in the early 1960s whose studies focused on the statistical analysis of communication networks and queuing theory, was another key player in the origins of packet switching. In 1961, he published the first paper on the theory of packet switching; and in 1964 he published the first book on packet switching, "Communication Nets." Davies, Roberts, and Kleinrock all met at the ACM Symposium. This meeting led Roberts to revise his ARPANET plans in order to integrate packet switching into the design of the network.

As described in chapter 2, the backbone of the ARPANET relied on packet-switching minicomputers called interface message processors (IMPs), which were interconnected by 56 Kbps lines. Since the network was in effect packaged into these IMPs, they functioned as both an interface between each host and the network and as a packet switch. In its role as a packet switch, each IMP received and transmitted packets, checked that they were intact and requested retransmission if they were not, and determined the best routing for each packet, based on continuously updated statistics stored within the IMP and provided by neighboring IMPs. It was, therefore, these dedicated IMP computers that best embodied the features and functionality of the first packet-switched network. They both formed and operated the network. Other, very different types of packet switching implementations were not far off, however. Moreover, as the ARPA engineers' understanding of networking increased and new demands were made on the network, their original design for packet switching would soon undergo considerable revisions in its implementation.

Theories about packet switching, especially those related to network performance and the behavior of the network as usage increased, could not compare to a working packet-switched network. No one knew precisely what to expect, because no one had ever before built and operated a packet-switched network. Engineers at ARPA understood, however, that they would need to make arrangements to closely monitor the network. Only through monitoring the network's operation and statistical analysis of its performance could they recognize the strengths and weaknesses of their design. They could then use that knowledge to adapt their design accordingly.

People at UCLA, the location of the first ARPANET node, offered to organize and run a Network Measurement Center under the supervision of Kleinrock. The behavior of the network would be monitored and studied; experiments would be performed to test the limits of the system; and statistical analysis would be conducted to better forecast how the network would perform in the future. It's hardly coincidental that some of the most prominent people in the history of the early Internet worked at the center, including: Vinton Cerf, one of the creators of TCP/IP; Stephen Crocker, creator of the Request For Comments document series, Jon Postel, one of the creators of DNS, and David Crocker, creator of some of the earliest standards for email. Watching and understanding the behavior of the first packet-switched network proved critical to the Internet's evolution and to an understanding of networking in general.

By the mid 1970s, ARPA was operating three separate and distinctly different networks, but at the core of each network was packet switching. Then came the Internet Program and the first interconnection of dissimilar networks into one integrated network, or Internet. As presented earlier, the project interconnected ARPA's point-to-point IMP-based network with the broadcast packet radio network of PRNET and the high speed satellite network of SATNET. Each of the three networks operated on different assumptions. The ARPANET, for instance, considered the network very reliable and assumed the IMPs were delivering packets in the correct sequence for them to be reassembled into the original message. The PRNET used radio frequencies to broadcast its packets and assumed that transmission could easily be disrupted by any number of sources. Packet switching on the

PRNET was in many respects the opposite of that on the ARPANET: it assumed the network to be unreliable. Accordingly, it was engineered with specific features to compensate for this lack of reliability. To make matters more complicated, especially with respect to packet switching, all three networks were built with different interfaces, packet sizes, labeling, and transmission rates. This meant that the ARPA engineers not only had to figure out how to get the networks to talk to one another, but that they also had to account for these differences in the packets themselves.

In only a few, short years packet switching had gone from theory to the first multi-node network to three separate and uniquely engineered packet-switched networks. Each of the three networks was a success and, if anything, proved the versatility and efficiency of packet-switched networking. But the ARPA engineers still had to consider how to reengineer their creations, including their implementation of packet switching, in order to interconnect them. Meanwhile, the military was also making their own special needs known with respect to packet switching. They not only wanted to interconnect different network types, but they also wanted to optimize each network for specific environmental needs. A successful military communications network needed to accommodate a wide variety of environmental conditions, and it needed to account for usage over land, sea, and air. Just as with the ARPANET, PRNET, and SATNET systems, packet switching would need to handle the general flow of traffic, but the specific technologies and implementations would differ. The underlying question, however, remained the same: how to go about interconnecting dissimilar networks and, while doing so, account for any number of differences with respect to packet switching implementations and network-specific expectations.[3]

While ARPA worked on how to engineer interconnecting the networks, and what these engineering changes might mean to their implementations of packet switching, other packet-switched networks were emerging and provided helpful comparisons for the efforts of the ARPA engineers. One such network was called Cyclades and was funded by the French government. It consisted of a much simpler packet switching system than that built for the ARPANET; and it had been designed from the start with the expectation of being interconnected with other networks. The Cyclades network delivered individual packets known as datagrams. These datagrams represented a more simplified form of packet, or message unit, than those designed for the ARPANET. Cyclades left all of the more complex functions — like those designed into the ARPANET's IMPs for checking errors and handling the sequencing of packets — to the hosts connected to the network. This allowed for a more streamlined network design and it reduced the amount of information that had to be contained in each datagram. At the same time, it meant that the bulk of the network intelligence resided in the host computers. For a while, the terms packet and datagram were used interchangeably by many. Both the ARPANET and Cyclades packets were self-contained and independent units of data. Both carried enough information to get them routed from the source computer to their destination. But, eventually, packet replaced datagram as the term of choice, with any subtle difference in definition being lost in the process.

The type of architecture used in the building of Cyclades had been considered but deliberately avoided by the ARPANET engineers. Taking into account the increased difficulties and the additional time it would have required to design, implement, and test a common set of communication protocols on each and every host that would be attached to the ARPANET, it was apparent that they would never have been able to get the network started if they hadn't instead designed the IMPs to consolidate the network operations and thereby minimize any impact or demands the network would make on the host computers. But circumstances had changed. It was becoming clear that adapting the host computers could no longer be avoided, especially given the task they now faced in determining how to interconnect the different networks and accommodate all the packet switching differences. Cyclades helped to clarify this point. Its simpler network design and dependence on host computer protocols provided the ARPA engineers with a framework on which to model the Internet.

Another type of packet-switched network being developed in the 1970s was designed from the start, like Cyclades, with the need to interconnect dissimilar networks; and, like Cyclades, this design consideration directly impacted the network's implementation of packets and packet switching. Xerox PARC researchers in the U.S. were developing their own proprietary packet-switched system called PARC Universal Packet (Pup). Its purpose was to interconnect several wide area networks, specifically the ARPANET, PRNET, and Xerox's company network, with their local area networks running Ethernet. Like Cyclades, Pup employed a simple datagram for the delivery of packets and relied on host-resident protocols to provide the intelligence in the network and to ensure reliable data transfer. It was the technical simplicity of Ethernet that was largely responsible for the design of Pup. Ethernet was the lowest common denominator in the network equation. An Ethernet network is simply a cable connecting host computers. There is no switch or router, so the hosts themselves essentially run the network. In Ethernet's network paradigm, the host computers are the network. Thanks to TCP/IP, this is also true of today's Internet, in which the packet switches focus entirely on moving the data from point to point as quickly as possible.[4]

It took only a few short years for these very different types of packet-switched networks to be engineered, built, and become operational. The ARPANET had been first, thanks largely to the creation of the IMPs. But the Cyclades and Pup networks provided one of the principal design changes that would allow ARPA to interconnect its three networks and, more importantly, be prepared to interconnect future networks. The IMPs would have to be modified or replaced and a new host protocol would have to be written to transfer responsibility for running the network to the host computers. This would take care of simplifying the network and would enable the network to grow organically as new types of computers connected to the Internet. The host protocol that was written was called the Transmission Control Protocol (TCP). Soon afterwards, it evolved into the protocol suite called TCP/IP. TCP, as well as TCP/IP, were written by ARPA engineers Vinton Cerf and Robert Kahn, and are discussed at length in the following chapter.

The other piece of the packet switching solution involved the physical connection between two networks. The ARPA engineers still needed a way to accommodate differences in such things as packet size and transmission speed. This is where specialized routers, initially called gateways, came into play. These devices were designed to direct packets from one network to the next, and to store information about each of the connected networks necessary to handle any differences in the local packet formats. This approach localized the differences between networks and, like transferring responsibility for the operation of the network from the network to the hosts, it greatly facilitated future network expansion. As described in chapter 2, ARPA conducted an Internet experiment in 1977 to show the U.S. Department of Defense what they could do with their newly connected Internet. They successfully sent packets on a 94,000 mile round trip across the ARPANET, PRNET, and SATNET. No data was lost in the transmission.

The new Internet paradigm, of a simplified network with host-based protocols and host-based network intelligence, meant that the network would now be considered unreliable at all times. For the ARPANET and Internet engineers this assumption was in large part made possible by packet switching. Packet switching kept the network simple, distributed, efficient, self-maintaining and self-repairing, thereby overcoming its potential unreliability. All the switching nodes were equal and they functioned independently to send and receive messages. Each message was divided into packets, and each packet was addressed and routed through the network separately.

Packet switching on the Internet can be summed up as follows. It is:

- **Digital**: Transmissions that were digital could travel across any number of connections and remain free of errors.

- **Computerized**: Switches that were themselves computers allowed the network to be monitored at each node, to be upgraded as needed, and even to repair itself.

- **Redundant**: The network would continue to function even if parts were damaged, and it could quickly accommodate fluctuations in usage.

- **Efficient**: Communication links were shared, carrying messages from any number of systems, and they only needed to be open for as long as data was being transmitted.

Even though the analogy to a telephone network may come more readily to mind, think of packet switching and the way it transmits data across the Internet instead as something more closely resembling the postal system. As with a letter delivered by the postal system, a packet delivered over the Internet contains a source address and a destination address, its delivery is guaranteed and passes through intermediate locations, it does not require that someone be there to receive it, and it travels with many similar items to save on transportation costs. Alternatively, consider the similarities between packet switching and a telegraph network. Each switching node in a packet-switched computer network functions very like a station in a telegraph system, copying down an incoming message and then transmitting it onward to the next node. Incidentally, telegraph offices in the 1960s began replacing people with computers to handle this same copy, store, and forward task; a fact that would have been known to the ARPANET engineers.

Although many originally doubted it or failed to recognize its importance, the ARPANET and the Internet clearly proved that a packet-switched network was better engineered to handle data communications than a dedicated circuit-switched network. It was also faster and less expensive to operate. Sharing the bandwidth provided by a packet-switched network makes today's Internet possible by keeping costs low and by allowing data of all types to pass through the network on an equal status with all other data. Among other things, this means that you don't have to worry about your short love letter getting stuck behind some endless treatise on computer networking. The division of all data into small packets before it gets passed onto the network for delivery ensures this. Today's Internet is the ultimate testament to the inventive, forward-looking work of both Baran and Davies. If it weren't for the creation of packet switching, the variety and pervasiveness of networking we enjoy today could never have come into existence.

# Protocols: The Definition of Interoperability

**5**

## The What and Why of Protocols

Have you ever wondered how computers are able to talk to one another, how an email message composed on your computer can be received and read on your friend's computer, how a Web page stored on some computer halfway around the world can be quickly and effortlessly displayed on your computer? The answers can all be found in an examination of the Internet protocols. Few, if any, people would consider protocols to be the most exciting part of the Internet. But if you want to understand how the Internet functions, you must understand the role they play and the place they occupy in the overall architecture of the Internet. You also need to understand, at least in very general terms, the types of operations that protocols perform and how they go about performing those operations.

Every basic service provided on the Internet, including sending and receiving email, browsing the Web, interacting with a newsgroup, logging on to a remote computer, or simply transferring a file, relies on one or more protocols for its operation. Each protocol consists of a set of requirements that precisely define the behavior and characteristics of a specific utility, like file transfer or the transmission of email. These requirements, or rules, which are similar to those in driving that specify that we must stop at a red light or yield to oncoming traffic, are incorporated by hardware, software, and operating system manufacturers into the products they build. The adherence to protocol requirements in these products ensures some measure of conformance to agreed upon standards, which in turn makes possible the fundamental interoperability that is the Internet.

When the ARPANET was being engineered during the late 1960s, protocols emerged as the means by which computers could interact with one another, something they never before had to do. More specifically, protocols defined how one computer could open a conversation with another computer across the network, exchange information with that computer, control and perform any number of highly specific operations, and finally close the conversation. At that time, the IMPs of the ARPANET (the minicomputers attached to each location's host computer that were built exclusively for networking) embodied the network, packaging, routing, and delivering the data across the network and ensuring the speed, safety, and reliability of the data's transmission. But the host computers (the large mainframe computers that held individual user accounts, stored data, ran applications, and performed calculations) had their own role to play. On one level, they had to interact with their local IMP to get data on and off the IMP subnet, something common to all the network-based operations they might be requested to perform. Whether they were asked to send an email or log in to another host computer, the associated data coming from the user had to be delivered to (and, at the far end, received from) the local IMP. On another level, each host computer had to interact directly with other hosts on the network. This host-to-host interaction, as opposed to the host-to-network interaction with an IMP, controlled the elements of negotiation and information exchange specific to each operation, such as presenting a login prompt during an operation for remote access or responding that access was granted or denied during an operation for a requested file transfer. All of these operations were defined and implemented in a collection of protocols that would reside on every host connected to the ARPANET.

The first host-resident protocol (the precursor to TCP/IP) was the Network Control Protocol (NCP). NCP packaged together the common network operations that each host computer would need to perform. These operations included all of a host's interactions with its local IMP, the low-level rules that controlled how it sent and received data for transmission across the IMP subnet. These operations also included the host-to-host functions common to all host-to-host interactions, such as initiating, maintaining, and terminating a connection. NCP, the network-layer protocol, would not be of much value, however, without application-layer protocols to define and implement the specific operations users wanted to perform across the network. Accordingly, the first application-layer protocols were created shortly after NCP: TELNET, for remotely logging into another computer, and the File Transfer Protocol (FTP), for transferring files from one computer to another.

Each of these protocols, NCP, TELNET, and FTP, like all the protocols that would follow, existed in two fundamental forms. One form was the definition itself: the specific requirements that described in clear and precise language all the operations a protocol needed to perform, and how it would negotiate those operations with another computer. A protocol might define, for instance, the sequence of steps that needed to occur for two hosts to say hello to initiate a conversation and to say goodbye to disconnect. The definition might specify that *EHLO* signified hello, and the computer signaling hello would then wait for one of several possible, predefined responses before it proceeded any further. What the protocol definition would not specify was how these requirements were to be implemented — what form the coding would take to program the operations to occur on any one type of computer.

Accordingly, the other form a protocol would take was its specific implementation on any one host. How a protocol was implemented, and even which features would or would not be implemented, was different for each host. While the implementation details varied, however, the operational details — what actually took place — needed to be the same for each host. Otherwise, the host computers would remain as isolated as they had been before the creation of the network. The protocols established a highly regimented, predefined framework that allowed dissimilar computers to talk to and interact with each other. As such, these first protocols, like all those that followed, formed the service-based building blocks of the Internet. As new services were created, they brought along with them new protocols, which defined each service and enabled implementations to be built for the various and sundry computer platforms that existed. The creation of the World Wide Web, for instance, brought us the HyperText Transfer Protocol (HTTP).

Today each and every host computer on the Internet must minimally implement the basic networking protocols of TCP/IP, which together provide the fundamental transport mechanism for exchanging data and the requirements for how to package data for transport over the Internet. Each host can then additionally implement as few or as many application-level protocols as desirable, all depending on how the computer will be used. Most computers these days are sold with a large and relatively standardized assortment of protocols that enables them to connect easily and immediately to a local network and to the Internet. Moreover, as you add hardware or software to your computer, you may also be adding and activating additional protocols as part of the process. For example, if you decide to add a wireless network card to your laptop, the installation will include the addition of one or more wireless protocols that are essential to the operation of this technology.

The implementation of any protocol will vary from computer to computer. Variations even exist between computers from the same manufacturer. These variations can be attributed to any number of factors, such as differences in the type of specific hardware components packaged with each computer, differences in the features or versions of hardware components, or differences in versions of the operating systems. Nearly all computers vary to some small or large degree with respect to their features and capabilities; and all are built with different compatibility requirements that affect whether or not they are capable of working with specific products, such as a proprietary application sold by the same company that manufactures the computer. All of these variations compete with the fundamental purpose of a protocol, which is to provide the means for different types of computers to interact and interoperate, to overcome their differences and find sufficient commonality through which to communicate and exchange information. Protocols are written, therefore, to accommodate such differences. As explained below, each protocol includes a core set of requirements that must be included in an implementation if it is to be considered compliant, while optional requirements provide room for additional features to be included. This is precisely what makes possible the interoperability of the Internet.

Protocols do not simply contain a list of unqualified, categorical requirements. They are actually far more refined, accommodating, and forward-looking in their approach to ensuring interoperability. The idea is not a lowest-common-denominator effort, sacrificing features and quality for the sake of some baseline uniformity. If anything, the goal is the reverse: building the best possible services while still achieving basic interoperability. Another working goal, often more implied than stated, is simplicity. These ends are achieved in part by dividing protocol requirements into one of three states, as follows:

- **Required**. Also labeled a **must** requirement, this type of protocol element is an absolute requirement.

- **Recommended**. Also labeled a **should** requirement, this type of protocol element has been recognized as one that may need to be ignored under certain conditions.

- **Optional**. Also labeled a **may** requirement, this type of protocol element may be included or excluded as determined by the marketplace for which the vendor is creating the implementation.

An implementation is said to be *compliant* only if it meets all of the **must** requirements. It is considered *unconditionally compliant* if it also meets all of the **should** requirements. It is considered *conditionally compliant* if it meets all the **must** but only some of the **should** requirements.

The different types of requirements and different levels of compliance provide a degree of latitude capable of accommodating some measure of flexibility in a protocol's features, the specifics of which typically arose during discussions of a protocol's implementation. This latitude was often critical to arriving at consensus among a wide assortment of corporations and organizations, both commercial and non-commercial, interested in the development and acceptance of any protocol. Parties interested in more advanced features for a particular service that could help them sell a product might try to get those features defined as optional. Other parties who had to contend with implementation limitations or backwards compatibility issues with their older product lines might argue for particular recommended requirements in order to maintain interoperability. Enabling a protocol definition to accommodate these different interests and needs while arriving at a baseline consensus to establish basic interoperability was key to the success of any protocol.

Simplicity was no less important in contributing to the success of a protocol than its ability to absorb some measure of variability in its implementation. As stated succinctly in the following excerpt from the Simple Mail Transfer Protocol (SMTP), a critical element impacting a protocol's acceptance and longevity is the simplicity of its design:

> SMTP's strength comes primarily from its simplicity. Experience with many protocols has shown that protocols with few options tend towards ubiquity, whereas protocols with many options tend towards obscurity.

The series of technical documents called Request For Comments (RFCs) have played an integral role in the development of protocols and their acceptance as Internet standards. They did not, however, start out with that objective in mind. RFCs originated as a modest and effective tool for documenting and debating technical issues during the early days of computer networking. The very first RFC (i.e., RFC 1), written by Steve Crocker in 1969 and entitled "Host Software," served to summarize work that had been done to date in ARPA's design and implementation of the ARPANET. Crocker uses the word protocol only once in the document. But what he is describing throughout the document are the simplest and most fundamental protocol requirements: tasks specific to the IMPs in performing their job of accepting, checking, and handing off data messages, or packets; and tasks specific to the hosts in performing their job of initiating, acknowledging, and maintaining a network connection. Every Internet protocol has been documented as an RFC. The series as a whole, therefore, records the history of the Internet's technology, defines what is required to produce compliant implementations of today's protocols, and serves to document the debates surrounding where the technology may be heading.

Two RFCs published in 1989 — RFC 1122 and RFC 1123 — describe the core set of protocols considered necessary for the operation of host computers on the Internet. These core protocols are no less important or prevalent today. RFC 1122 covers the protocols that reside in the communication layers, most importantly the TCP/IP protocol suite. As described earlier, these protocols define how hosts on the Internet communicate with one another and how data is turned into packets and transmitted across the Internet. RFC 1123 covers the protocols that reside in the application and support layers, such as FTP (used for basic file transfer), email, and DNS (which defines how host names are looked up and located on the Internet). These protocols define how we interact with the Internet by establishing the services that are available to us and, by extension, what we can and cannot do. Accordingly, the remainder of this chapter is divided into two sections: the first covers the communication-layer protocols and the second the application-layer and support protocols.

# The Communication-Layer Protocols

To understand how the Internet works, it's necessary to understand the role of TCP/IP, a suite of networking protocols that has been responsible for managing the transmission of data across the Internet since the early 1980s, and the single most important component making the Internet what it is today. If you think of the Internet as a global network of roadways, TCP/IP controls the flow and safety of traffic (i.e., the packets) along the roads and provides the vehicles that enclose and protect the contents of that traffic (i.e., your data) during its journey. Simply put, TCP/IP allows computers to communicate with each other regardless of their size, vendor, operating system, location, and a wide range of other factors. TCP/IP was able to achieve this baseline and widespread interoperability due largely to the fact that it was from its inception, and remains, an open, non-proprietary, deliberately generalized architecture with no predilection for any one type of computer hardware or operating system. Not only is the protocol fully documented and freely available, but many implementations, along with their source code, can be obtained for little or no charge.

If you use a computer today, any type of computer, more than likely it has the TCP/IP protocols already installed. No matter the chip, the manufacturer, or the operating system, no matter whether it sits at home, in the office, or in a computer center, even if it has never been connected to any type of network, your computer is almost certainly ready, willing, and able to function on the Internet. In fact, it would be difficult to purchase a computer today that is packaged without an implementation of TCP/IP. Yet there is no legislation mandating its inclusion. Instead, TCP/IP's pervasiveness is a testament to how far networking and the Internet have evolved, and to the critical significance of TCP/IP. It was not, of course, always like this, as the next section explains.

## The History of TCP/IP

The first ARPANET networking protocol was completed in December, 1970. It was named simply the Network Control Protocol (NCP) and was created by the Network Working Group (NWG), an ARPA-sponsored group made up of programmers from each of the four original ARPANET sites. NCP formed the middle layer in a simple, three-tier network architecture. It functioned to bridge the lower, data communications layer (composed of the IMPs and packet switching) and the higher, applications layer that handled direct user actions, such as those related to the copying of files or remote login.

NCP was designed to establish and maintain a communication link between two host computers. When an application called for data transfer to occur (like your using FTP to transfer a file), it would automatically and transparently make a call to NCP, which would then package the data into packets and forward the packets to the local IMP. Conversely, packets sent to the host from the local IMP would be accepted and handled by NCP before being forwarded along to the associated application (e.g., FTP or TELNET). NCP was specifically set apart in its own layer to simplify its implementation and to make it unnecessary for each application developer to write code for procedures that would be common to them all. Without NCP, for instance, every application would have needed to include its own code to handle the packaging and transmission of data, rather than simply handing off the job to a single, dedicated service as could be done with NCP.

Every host on the ARPANET had to have an implementation of NCP created for its specific architecture; and this had to be done before any applications could be written. All told, it took the better part of two years for all ARPANET hosts to have a working implementation of NCP. It did not, however, take long for limitations in the engineering behind NCP to emerge. Robert Kahn confronted the most significant problem while working on the packet radio system for ARPA that would become the PRNET. As discussed in previous chapters, packet radio was inherently unreliable. Signal jamming, general radio interference, disruptions and signal blackouts caused by tunnels or other obstructions all had to be accounted for. For a packet radio network to be viable, a reliable end-to-end network protocol was needed, one that would compensate for the inherent unreliability of the network itself by checking for and recovering from any breaks or failures in transmission of the data. But NCP included no such capabilities. It was written under the assumption that the network was already reliable, and any data transmission problems were handled elsewhere; specifically, it was expected that the network would monitor and resolve such problems, which is how the ARPANET was designed. For NCP, or any host-resident network protocol, to work for both the ARPANET and the PRNET, it would require altering part of the ARPANET's fundamental architecture and moving the responsibility for the network's functioning, the embodiment of the network itself, from the IMPs to the host computers. It would also require that significant changes be made to the existing protocols, especially NCP.

At first, Kahn considered developing a separate protocol for packet radio and continuing to use NCP for the ARPANET. This would have saved a considerable amount of development time and effort. But another limitation with NCP forced him to reconsider. NCP could only communicate with an IMP. It included no facility to address another host or another network. With other networks emerging, such as the PRNET and the satellite-based SATNET, and the growing interest from both the military and the academic community in interconnecting these dissimilar networks, this limitation had to be removed in order for the ARPANET to grow and evolve. Although its usefulness was short-lived, NCP had served its purpose. It effectively took the ARPANET from theory to working network. With networking itself only a few years old, however, the quickly emerging importance of internetworking had already become the guiding design principle for the future of networking.

Accordingly, Kahn set out to develop the replacement for NCP, which would eventually be called TCP/IP. The following four requirements were central to his development effort; and, not coincidentally, these same four requirements are part of the foundation upon which the Internet was built and are just as important to its operation today.

- Each network connecting to the Internet would be considered separate and self-contained; and no requirements would be made on any network for it to connect, enabling any existing or future networks to be accommodated.

- The network as a whole would be considered unreliable; if a packet failed to arrive at its destination, the protocol would be able to signal for its retransmission.

- Networks would be interconnected by *black boxes*: simple devices capable of translating network differences and unconcerned with the flow of packets. (These devices would later be named gateways; eventually they would be called routers.)

- No centralized or global control would exist; instead, the network's operation and control would be distributed throughout the network.

Other requirements for TCP/IP delineated by Kahn were as follows:

- Algorithms for successfully retransmitting lost packets and for preventing lost or damaged packets from overrunning and/or disabling the network.

- Host-to-host tunneling to allow groups of packets to travel together, provided the hosts and intermediate networks supported this feature.

- Gateway functions to forward packets, read headers for routing instructions, and adjust packet features (e.g., size and transmission speed) as necessary.

- End-to-end checksums for establishing data integrity, detecting duplicate packets, and reassembling packets from fragments.

- A methodology for global network addressing.

- Host-to-host flow control functions.

- An ability to account for operating system differences.

In the spring of 1973, Kahn began engineering work on the new protocol. He immediately teamed up with Vinton Cerf, whose knowledge of NCP and work with NCP implementations on various operating systems complimented Kahn's broader architectural understanding of what was needed. Their first paper described a single protocol, TCP, which included all services related to data transport and data forwarding. TCP purposefully supported the widest range of network types. This included the connection-oriented, virtual circuit type network exemplified by the ARPANET and its model of reliable, sequential packet delivery and the simpler, connectionless, datagram service type network perfect for a network as unreliable as packet radio, in which packets might be lost, delivered out of sequence, or corrupted.

Early experimentation made clear the need to divide TCP into two separate protocols in order to meet the needs of such diverse networking models. Early packet radio work, for instance, showed that it was necessary for packet forwarding and packet routing to be handled at the transport control layer, while packet loss issues could be handled at the higher, application layer. Consequently, the simpler services of forwarding and routing packets were bundled into IP, and TCP was pared down to provide the end-to-end services of flow control, packet sequencing, and the retransmission of lost or damaged packets. An additional protocol, called the User Datagram Protocol (UDP), was also created. It functioned as a smaller, simpler version of TCP and provided the necessary access to IP for applications that did not want or need all the services of TCP, such as its costly reliability and retransmission services. Good examples of applications that typically use UDP include radio broadcasts and games over the Internet. In these applications, the retransmission of dropped packets (a principal element of TCP) would do more harm than good.

Separating out IP provided additional benefits, like simpler, more efficient routers that no longer needed to duplicate TCP functions handled by the hosts. In turn, simpler routers lessened the general assumptions about, and the demands on, interconnecting networks. Requirements for network-to-network interaction could now focus exclusively on the packets, with the routers localizing any necessary changes to the packets' size or format and remaining unconcerned with the packets' flow, sequencing, or even if they had become damaged. Another benefit was that military networks could also be made simpler and more robust, while separating out IP provided a new plug-and-play capability that allowed new network types to be added without causing any disruptions to overall service.[1]

It's interesting to note that the first TCP model was conceived from the perspective of large-scale networks that spanned nations, like the ARPANET, something very different than what came to pass and what we find with the Internet today. But this was understandable given the existing conditions and expectations. Networking was still in its infancy, and the notion that thousands or millions of individual networks would come into existence in the near future, or ever, was simply unthinkable. Ethernet (the current de facto standard for creating small networks) was still in its development stage at Xerox PARC, and the emergence of local area networks (LANs) was not yet on the horizon. The early model of TCP, therefore, used only one-quarter of an IP address (i.e., 8 bits of the total 32 bits available) for network identification and the remaining three-quarters (i.e., 24 bits) for host identification. This meant that there could be a maximum of 256 networks. When LANs started to become a prominent feature of the network landscape in the late 1970s, it became painfully clear just how wrong the original assumptions had been. Accordingly, the IP address model was inverted, with 24 bits representing the network and 8 bits representing the host. Moreover, the new network address model was further divided into three classes, with the largest — Class A — representing networks of national scale, the next — Class B — representing networks of regional scale, and the smallest — Class C — representing local area networks.

Every host computer required an implementation of TCP/IP. The work of producing these implementations was originally funded by ARPA. The first versions of TCP/IP were created for large, time-sharing systems. But as workstations and personal computers came on the market, versions of TCP/IP suitable for these architectures followed quickly. Perhaps the most significant implementation of TCP/IP was the one funded by the University of California at Berkeley for a version of the Unix operating system. Berkeley's incorporation of TCP/IP into the Berkeley Software Distribution (BSD) Unix system in 1983 brought networking and the Internet into much of the computer science research community throughout the U.S. This, in turn, increased interest in and the number of applications for the Internet.

Incidentally, Berkeley's funding of a TCP/IP implementation for BSD Unix led indirectly to the creation of Sun Microsystems by Bill Joy. As a graduate student at Berkeley in the late 1970s, Joy worked on several releases of BSD Unix and then worked on the ARPA contract for developing TCP/IP. Not long afterwards, Joy founded Sun and began producing the first minicomputers running version 4.2 of the BSD Unix operating system. Complete with TCP/IP, they were shipped ready to plug into the Internet.

The most significant date in the history of TCP/IP was January 1, 1983. This was the date specified as the cutover point for migrating all the ARPANET hosts from NCP to TCP/IP. The transition was planned over several years leading up to the cutover deadline; and details about what was required and what would happen should you fail to comply were well communicated throughout the ARPANET community. Seventeen years later the Y2K transition would consume billions of dollars and create a media frenzy. But in terms of technical challenges, the Y2K work was trivial in comparison.

The first step of the transition took place in January, 1981 when the existing Internet packet format was modified from 32-bit headers to 96-bit headers. (IP headers are described below. Essentially a header is a predefined segment of each packet that carries information describing the data contained in the packet: its size, where it's going, where it's coming from, and so on.) At that point, network applications residing on each host also had to be adapted to use the new format or they would no longer be able to use the ARPANET. The second step of the transition was the migration from NCP to TCP. This included updating the FTP and TELNET protocols and incorporating a new addressing scheme for mail and a new mail protocol called the Simple Mail Transfer Protocol (SMTP). Local IMPs and TIPs also had to be replaced with new models from BBN that ran the new Internet protocols.[2]

About half of the ARPANET sites met the deadline, mostly through long hours of hard work and careful planning. Others, however, did not take the threat of being disconnected seriously or left the work till the last minute without understanding how long it would take. It took another six months for the transition to TCP/IP to be completed. All that remains of the event now are buttons that read: "I survived the TCP/IP transition." It's hard to imagine such a massive, synchronized technology change occurring on the Internet today.

What made the 1983 transition to TCP/IP possible was that the U.S. military adopted TCP/IP as a defense standard in 1980; and it was the military that owned and operated the ARPANET. More to the point, the military was making ever greater use of the ARPANET now that the networking technologies were proven and interconnecting different types of networks was possible. But the majority of ARPANET hosts were at non-military research institutions, and control over access to and use of the network was not sufficient to maintain the kind of security needed by the military. The cutover to TCP/IP, therefore, also enabled the Department of Defense to split the ARPANET, creating the MILNET to support the military's operational requirements and leaving what remained of the ARPANET to support continued research.

In 1985, the National Science Foundation (NSF) announced its decision to adopt TCP/IP for its NSFNET program, which was being built to provide a wide area networking infrastructure for facilities of higher education throughout the U.S. As discussed in earlier chapters, the NSFNET provided a path for commercialization of the Internet and ended up assuming responsibilities for the Internet backbone when the ARPANET was decommissioned in 1990. NSF's decision to mandate the use of TCP/IP was key to facilitating this transition.

TCP/IP was, however, far from being the only choice at the time. Networking protocols other than TCP/IP were developed and widely used during the 1980s. Some, like TCP/IP, were open source and were distributed freely. Open-source protocols were responsible, for example, for bringing many popular, grassroots networks into operation, like BITNET and USENET. Other networking protocols were proprietary in nature and were developed in the commercial sector to provide for more specialized and localized networking needs. These included XNS from Xerox, DECNET from Digital, and SNA from IBM.

These same commercial suppliers of proprietary networking products also developed commercial, TCP/IP-based products for their customers to use with the Internet. Most of these computer companies included the Department of Defense as one of their largest, most important customers; and the Department of Defense stipulated in their contracts a requirement for TCP/IP. A growing complaint from these corporations at the time was that there was not much information available, from the government or elsewhere, on how the Internet was being used or much guidance as to what products were needed. As a result, a three-day workshop was organized in 1985 in cooperation with the Internet Activities Board (IAB) that brought together 250 vendor representatives and 50 inventors and experimenters to discuss what TCP/IP did and what it could not do. The exchange of information generated interest and excitement with both the commercial vendors and the product demonstrators, and marked the beginning of a discussion that would last more than ten years.

Two years of conferences, tutorials, and workshops followed the initial meeting and culminated in September 1988 with the first Interop trade show. Fifty companies along with 5,000 engineers attended the event, which successfully demonstrated both the widespread commercial integration of TCP/IP into a large variety of new products and the all important interoperability of the protocols, even between competitors' products. The Interop trade show has since grown substantially, with meetings held in seven locations worldwide and an audience exceeding 250,000 people.

By 1990, it was clear to most people involved that the race (if it ever qualified as a race) for dominance in the area of networking protocols was over. There was no arguing against the fact that TCP/IP had become a de facto standard and, in the process, had eliminated or diminished the significance of most, if not all, of its competition. Moreover, this was true not just in the U.S., but worldwide. The Internet is the best proof of this dominance, and the best indicator of TCP/IP's overall importance. As presented in chapter 3, even the 1995 resolution by the Federal Networking Council to provide a working definition for the Internet relies on TCP/IP, referencing both the network addressing mechanism that is part of IP and the communication services provided by both TCP and IP. Nothing could better convey how fundamental TCP/IP was to the nature of the Internet, and, indeed, to its very existence.

## The TCP/IP Model

Think of the process of networking as data progressing through a series of layers, with each layer stacked one on top of the next like the ingredients of a sandwich. This process is responsible for converting your actions (e.g., typing letters at a keyboard or clicking a mouse button to make a selection) into the data that the attached network accepts as traffic. As discussed in chapter 3, the OSI model arranges network functions into a hierarchy of seven layers, each representing a different aspect of the communication process. For our purposes, we only need to examine four of these layers, as follows:

| TCP/IP Protocol Layers | |
| --- | --- |
| Communication Layers | Protocols |
| Application | TELNET, FTP, Mail, etc. |
| Transport | TCP, UDP |
| Network | IP, ICMP, IGMP |
| Link | Network Interface and Device Driver |

TCP/IP forms the middle layers in this network sandwich, the ham and cheese between the outer layers of bread, with your browser or email application as the top piece of bread and your computer's network card as the bottom piece. Together, TCP and IP form a bridge between your interaction with the computer, when you hit the *Send* button to dispatch an email, and the computer's interaction with the connected network, when it transmits your mail's data along with its destination and return address information across the computer's network interface card out onto the network. In general, data is sent down the protocol stack, traversing each layer, until it is transmitted as a stream of bits (zeros and ones) across the attached network. In the process, each layer handles and packages the data according to its specific needs, as shown here:

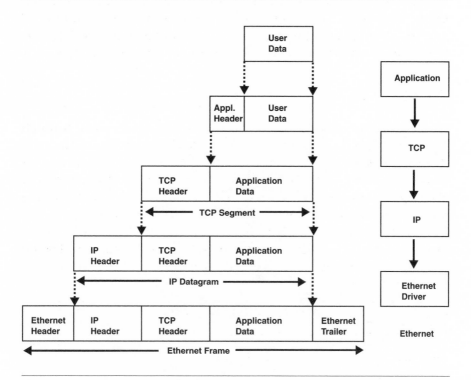

**Figure 1.   Network Layers Model**

Note that, as the data from the application makes its way down the stack, each layer holds progressively more information to describe it. The process begins when an application like FTP calls on TCP to transmit its data. The application passes along buffers (short bursts) of its data to TCP. TCP then packages these buffers of data into its own units, or segments, and calls on IP to accept each segment and forward it to the destination host's TCP layer. At the destination, once the IP packets have made their way across the network, the receiving host's TCP layer reverses the process by taking the segments of data and passing them along to the receiving application process. Most importantly, the control information embedded in each layer's separate and unique header area ensures the reliability of the transmission.

When it comes to the data traversing the network, TCP depends on IP to provide the interface to the local network and to get the packets across the network as quickly and as efficiently as possible. IP performs its own repackaging of the data segments it receives by encapsulating them inside datagrams that it then routes directly to the destination or to the local gateway, depending on where the destination is located. Packet switches along the way may effect further operations on the data in order to facilitate delivery, depending on where the data is going and on any differences among the intervening networks along the journey.

Out on the network, a router functioning as a gateway between two networks unwraps each IP datagram, examines its header information to determine which network it must travel to next, re-wraps the datagram in that network's local packet format, and sends it on its way. If necessary, a gateway may divide the datagram into smaller fragments; and this may happen any number of times along the datagram's journey. Information is supplied with each repackaging so that the original datagram can be reassembled from these fragments at its destination. This is how and where different networks that interconnect to form the Internet both route your data through the Internet and automatically adapt the packets to accommodate network differences.

When data is received at the destination it travels up through the network layers. The packaged headers are read, removed, and used to forward the data onward and upward. The destination IP reconstitutes the original data segments from the incoming datagrams and passes them on to the destination TCP. The transmitted data, once collected and restored, is then passed on to the associated application.

This entire process, passing your data down through the protocol layers on your computer, through TCP and IP and then onto and through the network, and then passing your data back up through the corresponding protocol layers on the remote computer, typically occurs in far less time than it takes to describe it. Not only is the process fast, but it is reliable. This is the true power and elegance of the Internet's architecture: it provides fast, efficient, reliable, adaptable, and transparent data communication. Some of the engineering details that make this possible are described in the following sections.

## The Transmission Control Protocol: TCP

As shown above, TCP resides in the middle transport layer of the four-tier network architecture. It interfaces with (i.e., bridges the gap between) the application layer above and the network layer below — the packet/messaging component that for our purposes will always be IP. TCP's primary function is to provide a logical, or virtual, circuit between two processes. Processes are defined as any computer elements or agents that read and write messages. This is an important point to grasp: it is with and through TCP that your computer and a remote computer on the Internet are effectively joined. Think of TCP as a telephone handset on each computer that is used to exchange your data and any associated control information as you copy a file or view a Web page. TCP is a control freak: its concern is handling the data exchange and making sure that no data is lost or damaged in the process. IP, on the other hand, is an adrenaline junkie; it just wants to get your data across the network as fast and as efficiently as possible.

TCP is a *connection-oriented* service: in order for data to be transmitted, TCP must first establish, and then maintain, a connection between the originating application and the destination (think of your Web browser and a remote Web server). The operation that establishes the connection is typically referred to as a handshake, because each end of the connection is acknowledging the other. The handshake is actually a three way affair, because the sender must also acknowledge the destination's reply. This process of acknowledgment helps to prevent duplicate connections. The connection itself consists of two TCP processes — the sending TCP on the originating computing and the receiving TCP on the destination computer — and it is through these two processes that the computers are joined and communicate over the Internet.

The virtual circuit established by TCP must be reliable and secure, despite the fact that the underlying service, IP, is inherently unreliable. To achieve the necessary level of reliability and security, TCP provides the following operational areas and controls:

- **Basic Data Transfer**. A continuous stream of data in the form of bytes (bytes are small groupings of typically — but not always — 8 bits) is provided in each direction and packaged into segments. Data is blocked, forwarded, and pushed (promptly or forcibly delivered) as conditions dictate.

- **Reliability**. Each transmitted data byte is assigned a *sequence number* and includes a checksum value indicating its size. The receiving TCP must acknowledge receipt within a certain time interval, otherwise the data is resent. The receiving TCP uses the sequence numbers to reassemble the segments and eliminate duplicates; and it uses the checksum value to identify damaged packets, which are then discarded.

- **Flow Control**. The receiving TCP controls the flow of data by including with its acknowledgments a *window* that shows the number of the last segment received and an acceptable range of sequences that can now be sent.

- **Multiplexing**. TCP provides multiple ports within each host so that TCP facilities can be used simultaneously by several processes. A *socket* is formed to uniquely identify each port through a combination of a port's identification with the host and network addresses of the TCP process; each socket can be used in multiple connections.

- **Connections**. For two processes to transfer data, they must initialize a connection using the handshake mechanism described above, maintain the connection using the flow control mechanism described above, and terminate the connection when they are finished in order to free system resources.

- **Precedence and Security**. Users have the option to embed optional information through TCP that can specify factors of security or priority related to the packaging and delivery of their data, respectively. Such features were requested for application in military communication networks.

Once a handshake occurs and a connection is established, a stream of data in the form of bytes is transmitted back and forth over the connection. The contents of the data stream are irrelevant to TCP. Its concern is ensuring safe delivery; the task of interpreting the data is handled by the application layer.

Each TCP data segment carries with it a TCP-specific header that contains the information necessary to establish the TCP connection and to ensure reliable delivery of the data. The IP service creates its own distinct header — as explained in the next section — to pass along information specific to its unique tasks. Incidentally, it is this storage of information in separate, well defined headers that describe both the data and its delivery that makes it possible to use host protocols other than TCP, if desired, and network protocols other than IP. The TCP header layout is organized as follows:

| TCP Header Layout | | | |
|---|---|---|---|
| Source Port | | | Destination Port |
| Sequence Number | | | |
| Acknowledgment Number | | | |
| Data Offset | Reserved | Control Bits | Window |
| Checksum | | | Urgent Pointer |
| Options | | | Padding |

All the information contained in a TCP header — or an IP header for that matter — is metadata; it is information that describes your application data as well as your computer, the destination computer, and the data transmission itself. So if you were ever wondering where the information is stored that gets your data across the Internet safely, efficiently, and quickly, the answer is, in these headers. The information stored in each area of the TCP header is briefly described in the following list:

- **Source and Destination Ports**. These areas hold the port numbers of the initiating host and the receiver host. These numbers, combined with the IP addresses contained in the IP header, uniquely identify the connection between the hosts, which is often referred to as a socket.

- **Sequence Number**. This area identifies the first byte of data (i.e., the beginning of the segment) in this segment of the data.

- **Acknowledgment Number**. This area identifies the byte of data that the sender should transmit next; the number is typically the sequence number of the last byte received plus one.

- **Data Offset**. This area holds the size of the header, which is necessary to locate where the header ends and the data begins.

- **Reserved**. This area is reserved for future use.

- **Control Bits**. This area holds six flags (called notations) that can be used together or alone to fine tune the data transmission.

- **Window**. This area indicates the number of bytes of data that the receiving TCP is willing to accept in the next exchange.

- **Checksum**. This area holds the calculated size of the TCP header and data when it was created; it is used by the receiving TCP to verify that the segment was not damaged during transmission.

- **Urgent Pointer**. This area holds an offset number that is used by the sending TCP to locate the last byte of urgent data as part of an effort to transmit emergency data to the receiving TCP. This pointer might be used, for instance, in a file transfer program when the user interrupts the transfer for some reason and TCP needs to notify the other end of the connection immediately, interrupting the normal flow of data.

- **Options**. This area holds optional information that needs to be communicated. It is most commonly use to specify the maximum size for a TCP segment.

- **Padding**. This area is a string (sequence) of zeros that pads out the header to a specific boundary.

TCP traffic includes a wide spectrum of applications, including email, file transfer, remote login, and browsing the Web. The traffic itself, however, can be neatly divided into two categories: traffic that is synchronous (or interactive), which requires an immediate connection between what you are doing and what is being returned by the far end of your connection (e.g., a remote login session) and traffic that is asynchronous, which does not require an immediate give and take (e.g., email or browsing the Web). TCP behaves differently depending on the type of interaction required by your application.

In a remote login session, TCP might well send each keystroke you enter as a separate TCP data segment. In this case, one byte of data generated by pressing a single key will become a data packet of 41 bytes, because the data packet must include 20 bytes for the TCP header and another 20 bytes for the IP header. This type of communication (consisting of many, very small packets) will not cause a problem on a local area network, where interaction between the hosts is likely to be very fast. But it might be an issue when logging on to a remote computer many network hops away, on a wide area network like the Internet. TCP handles such differences by providing algorithms and delayed acknowledgments that alter its behavior, based on its reading of network conditions and other programmed factors. Application programmers have access to some of these features so that they can choose which TCP features should and should not be used by the programs they design and build. For instance, allowing a slow connection to send a series of keystrokes together — rather than one at a time — to speed up data transmission would probably work well for a remote login session, but it would prove disastrous for an application relying on a user's mouse movements to capture real-time actions.

In a file transfer operation, TCP is generally transferring large blocks of data one way along a connection. While some things that affect the speed of this transfer are beyond the control of TCP (e.g., the quality and type of the network connections), TCP can adjust the the window size, which determines how much data can be transmitted in a segment, to maximize the flow of data based on other specified or discernible parameters of the connection. Each TCP connection is unique, and the protocol has the means to adjust to network conditions and to specific application requirements as needed. This is yet another aspect of TCP that illustrates its inherent adaptability, an important feature considering how rapidly the Internet and its technologies continue to evolve.

## The Internet Protocol: IP

IP datagrams, or packets, are the blood cells of the Internet's circulatory system. IP itself is a delivery service for these datagrams. As we have seen, IP was purposefully defined with expectations that its service would be unreliable. It makes a best-effort attempt at delivery, but it was designed not to slow down or become adversely affected in some other manner when problems occur, such as in the event of a power failure or hardware problem that results in datagrams being damaged or destroyed while traversing the Internet. If a failure occurs, IP will discard the affected datagram and attempt to signal the source that the datagram was lost so that it may be resent. But that constitutes the extent of its efforts. IP presumes that repeated or additional efforts to recover from the failure will be handled at a higher level, such as by TCP. IP's single-minded focus on delivering data is precisely what keeps data moving quickly across the Internet and prevents the network from developing bottlenecks.

Unlike TCP, IP is a *connectionless* or *stateless* service; each datagram is handled separately, independent of any datagrams arriving before or after it. This can lead to datagrams being delivered out of sequence, because different datagrams may take different routes to their destination. IP's job is the fastest and simplest delivery of the datagrams. Remaining oblivious to datagram sequencing, like not being responsible for lost packets, keeps IP running efficiently.

The IP datagram header layout is organized as follows:

| IP Datagram Header Layout | | | | |
|---|---|---|---|---|
| Version | Header Length | Type of Service | Total Length | |
| Identification | | | Flags | Fragment Offset |
| Time To Live | | Protocol | Header Checksum | |
| Source Address | | | | |
| Destination Address | | | | |
| Options | | | Padding | |

The IP header doesn't look all that different from the TCP header. The overall concept is the same; only the specific types of information differ. The information stored in each area of the IP header is briefly described in the following list:

- **Version**. This area holds the internet header format version number. Version 4, referred to as IPv4, is currently the most common. But the newer version 6, IPv6, is catching up.

- **Header Length**. This area holds the number of 32-bit words contained in the header.

- **Type of Service**. This area is used to specify the quality-of-service desired as related to the type of operation being performed with the data. It is here that networking characteristics, like transmission delay, reliability, and throughput, can be adjusted and balanced with respect to the type of data being carried. For instance, a file transfer operation might choose to maximize throughput, while a remote login operation might elect to minimize any delays.

- **Total Length**. This area holds the total length of the IP datagram in bytes. Used in conjunction with the header length, this value allows IP to determine precisely where the header ends, where the data begins, and the size of the data portion of the header.

- **Identification**. This area uniquely identifies each datagram and is used at the destination to aid in reassembly and in handling any fragmentation that might have occurred.

- **Flags**. This area contains a control value related to fragmentation of the datagram.

- **Fragment Offset**. This area indicates where this fragment belongs with respect to the parent datagram.

- **Time to Live**. This area indicates, in seconds, the longest time the datagram should be handled by the network. Zero indicates that the datagram should be destroyed. Each time the datagram is handled, this value is decreased by a minimum value of one, which represents the smallest increment, or one second. In simplest terms, this feature prevents datagrams from traveling the Internet indefinitely. It is also used to make sure that time sensitive data is delivered quickly or not at all.

- **Protocol**. This area indicates, using an assigned number, which protocol (e.g., TCP) will be used to hand off the data.

- **Header Checksum**. This area contains a computed value, related solely to the header component, used at each hand-off to ensure the integrity of the datagram. This value must be recalculated at each hand-off, because some header values (e.g., time to live) change. It specifically excludes the data, because that checksum is provided at the higher-level protocol (e.g., TCP).

- **Source IP Address**. This area contains the sender's IP address.

- **Destination IP Address**. This area contains the intended recipient's IP address.

- **Options**. This optional area can contain information on security and handling restrictions (designed for military networks), a record of the datagram's routing, a loose source routing (i.e., a suggested path), and a strict source routing (i.e., a pre-determined path).

IP routes data quickly and simply. If the destination is local (i.e., either directly connected to the sender or on the local network), IP datagrams are sent directly to the destination. Otherwise, the datagrams are sent to the network's default router, which connects the local network to a wider area network. The router then begins the process of sending the datagrams onward towards their destination.

IP routing consists of a series of hops. Routing tables that contain such things as a history of destination IP addresses and the IP address of the next router or attached network assist in the process of quickly forwarding datagrams on their way. Except in the case of a directly connected host, the complete route from sender to destination is neither known nor relevant to IP. Not knowing the route is key to the success of IP; all that matters to IP is that the next hop brings the datagram closer to its destination. For a datagram to continue on its journey, IP must find in the routing table an entry that matches the destination IP address, an entry that matches the network of the destination, or an entry labeled *default.* Otherwise, the datagram is considered undeliverable, and its journey ends then and there.

## The Future of TCP/IP

As the Internet, and networking in general, expands its reach, we can expect to find TCP/IP built into more and more of the products we buy, and not just computers. Mobile and wireless products are reshaping the technology landscape, enhancing the features of existing products like laptops and creating whole new product lines. Devices that merge and connect disparate technologies, like a cell phone that incorporates a digital camera and Web browser, all seem to have one thing in common: they are Internet-ready, which means they have incorporated TCP/IP into their architecture. It won't be long before TCP/IP invades other, less obvious products. Home appliances, like your refrigerator, dishwasher, and washer/dryer are potential future candidates. It is already possible to use the Internet for telephone calls, and it is expected that this service will become more widely used in the next few years. The next step will likely be Internet-ready phones, which suggests that Internet-ready radios and televisions cannot be far behind.

The expansion of networking owes its existence to protocols devised and defined in the late 1970s and early 1980s, when most people, if they had even heard of a computer, knew only what they saw on television or in the movies. Few knew anything about computer networking. Today, computers, computer networks, and the Internet seem to be everywhere. They are in our homes, at work, in public libraries and other public facilities (e.g., airports and civic centers), and routinely carried by people from place to place. TCP/IP is adapting along with the times, or at least it is trying to. As described below, the next generation of IP is already here, even if the majority of traffic is still being handled by the older protocol definition. The enhancement of TCP, on the other hand, remains largely mired in the discussion and debate phase.

The Internet Engineering Task Force (IETF) is responsible for the continued development of the TCP/IP protocol suite and any standards work related to it. The IETF was formed in 1986 when commercial TCP/IP products first started to be produced. They were, at that time, funded by the U.S. government and attended by a few hundred interested engineers from universities and research institutions. These same meetings are now attended by well over a thousand people. Costs are covered by user fees, and the majority of attendees represent commercial vendors. The overall objective of the meetings, however, has remained the same: to collectively evolve the TCP/IP suite in a cooperative way that represents the interests of the vendors, the end users, and the community at large.

## IP: The Next Generation

The IETF spent four years developing the standard for the next version of IP, called IP version 6 (IPv6) and known as IP Next Generation. The IPv6 protocols became a draft standard in 1998. The driving force behind the new version is that IPv4, the IP version currently controlling most Internet traffic, allows only 32 bits for an IP address and, therefore, allows for a maximum of roughly 4 billion addresses. (The 32 bits are composed of 4 separate numbers, each ranging from 0 to 255; for example, 192.168.0.1.) This may seem more than adequate to meet the need, but for two very simple reasons it is insufficient. The first is that the rate of growth on the Internet continues to be exponential, which means that it is only a matter of time before all the addresses are allocated. The second is that, because organizations acquire blocks of addresses based on their size and their projected needs and most planners overestimate those needs to allow for future growth, a sizable percentage of the maximum number of addresses will never be used.

IPv6 was designed to operate over the latest networks (e.g., high speed Gigabit Ethernet and the lower bandwidth wireless networks) and to be able to handle the newest emerging networking technologies. In addition to allowing longer addresses, IPv6 takes advantage of the latest high-end computers that use 64-bit processing and includes more sophisticated security features. The most versatile feature of IPv6 is that it automates how computers are identified on the network. In effect, it automatically assigns an address and configures other network parameters when a computer plugs into the network. With IPv4, addresses are assigned manually by system administrators, or they are dynamically allocated through a network server from a pool of available addresses.

The key features of IPv6 include the following:

- **Scalability**. IP address space is increased to 128 bits in length from 32 bits. This expansion equates to a maximum of 3x10^38 (3 times 10 to the 38th power) addresses. This is, simply put, a huge number. IPv6 also allows for more levels in the address hierarchy.

- **Security**. Integrated into IP in response to increasing privacy and security concerns are extensions that support authentication, data integrity, and data confidentiality.

- **Real-Time Support**. A new flow labeling mechanism can be used to identify packets for special handling. An application can use this feature to request a quality of service accordant to its needs, which might include facilitating real-time traffic for video conferencing.

- **Auto Configuration**. IPv6 can automatically configure host addresses, simplifying administration and address allocation.

Vendors must now support the new version of IP and integrate it into the computer products we buy, the computers that supply services on the Internet, and the routers that handle the internet traffic. The IETF has created a working group to oversee the transition; its specification provides a migration path that allows IPv4 equipment to continue functioning from an IPv6 address through a process called tunneling, in which an IPv4 header is placed on an IPv6 packet. Many manufacturers have already created IPv6 implementations of their products. The transition is meant to be slow and incremental and was designed with few interdependencies. This will allow both network operators and users to migrate to IPv6 without any coordination or disruption in service.

## TCP: The Next Generation

TCP, unlike IP, has not yet undergone a major revision, even though the debate about its need to adapt has been going on for several years. In the mid 1990s, the IETF formed a working group called TCP Next Generation (TCPng) to address how TCP might need to change in accordance with new networking facilities, equipment, and trends, such as the expansion of wireless computing. But the working group has not yet issued any standards or RFCs.

The existing TCP specification is optimized for wired networks. Networks that include wireless segments or nodes exhibit different characteristics than wired networks in the transmission of data, the anticipation of delays, and the handling of errors. For example, TCP interprets missing data segments as an indicator of network congestion and responds by signaling the sender to slow its transmission rate. For wired networks, this is reasonable. For a network that includes wireless nodes, this assumption could be inaccurate and the response could result in degraded performance.

Energy consumption is another area that warrants changes to TCP. Newer technologies like laptops and personal digital assistants (PDAs) rely on a limited battery supply for their power. But TCP includes no facility for adjusting its behavior based on the device running it. It presumes unlimited power is available and, therefore, wastes energy when it could be configured to conserve power, based on a set of conditions.

When the next TCP specification will be written is impossible to determine. But it's safe to say that wireless, mobile communication will play an important role in how it evolves.

# The Application-Layer and Support Protocols

The application layer, the topmost communication layer, represents the level at which we interact with a computer. Unlike the transport and network layers, which contain protocols that define the transmission and packaging of data (e.g., TCP and IP), or the link layer, which concerns itself with the physical transmission of digital signals that represent the data communication, the application layer contains the protocols that define which actions we can and cannot perform on some remote computer across the network, and precisely how those actions are communicated, evaluated, negotiated, and carried out between the two computers. The following core application protocols, which are described in RFC 1123, define many of our most basic and commonplace interactions with the Internet:

- Remote login: the TELNET protocol for connecting to and working on remote computers.

- File transfer: the File Transfer Protocol (FTP) for exchanging files with remote computers.

- Electronic mail: the Simple Mail Transfer Protocol (SMTP) for exchanging email with remote computer users.

RFC 1123 also describes the protocols that constitute the core support services on the Internet:

- Domain name translation: locating the IP addresses of remote computers.

- Host initialization: using the network to configure a host's network settings and resources.

■ Remote computer management: using the network to monitor and maintain computers from a remote location.

The protocols that handle these support services are less directly related to the actions we perform on the Internet than the application-layer protocols. But they handle functions that the other services rely on, such as translating a host's domain name into its corresponding IP address; and they provide functions that are important to the management of networked computers, such as the host management operations performed in data centers and in large-scale commercial networks, operations that will likely become more commonly used by other types of networks in the near future. Before discussing the application-layer and support protocols in more detail, the following section presents some of the design considerations found to be important to the success of any host-resident protocol, as expressed by the vendors of Internet host software and derived from their experience in implementing protocol specifications.

## Protocol Design Considerations

The protocols described in RFC 1123 are the building blocks with which the core services of the Internet were built, and which many subsequent services successfully leveraged, referenced, or incorporated. These core services fulfilled the requirements of the earliest networking projects, such as accessing a remote computer and copying files from one computer to another. Even today these services continue to account for a large percentage of the Internet's traffic. RFC 1123 was written just prior to the advent of the Web and the creation of its primary protocol, the HyperText Transfer Protocol (HTTP). The Web greatly popularized the Internet not by reinventing it, but by making its existing core services, like FTP and TELNET, much easier to use, and by building on other elements of the Internet's foundation. It's important to note, therefore, that without these core Internet protocols, the Web — and many other services — could never have come into existence.

Protocols are first and foremost about interoperability. This means that they must ensure that, despite any number of differences among the types, makes, and capabilities of computers across the Internet, any one computer can rely on certain basic functions being available and operating consistently on a remote computer accessible across the network. Accordingly, anyone seeking to contribute to the specification of a protocol or create an implementation should be careful to take the following general observations into consideration:

- The Internet is constantly evolving.

- Good software engineering is a matter of robustness (i.e., anticipating and planning for all eventualities).

As the history of the Internet makes abundantly clear, expectations concerning where the Internet is heading often run contrary to actual use. Changes are inevitable, and predicting what form they will take and the extent of their impact is impossible. Given the size of the Internet today and its continuing exponential growth, it seems only prudent to assume that problems will occur and that new hardware and software products, as well as innovations in networking itself, will alter the Internet's continued evolution. These factors will eventually demand changes to the core Internet protocols, and to any number of other protocols. There already exists a sizable history of changes to these core protocols resulting from a wide array of issues, including original engineering flaws that triggered data traffic jams and evolutions in hardware technology that called for corresponding protocol enhancements. Consequently, developing implementations of these protocols must be considered an ongoing process if the wide-scale interoperability of the Internet is to be respected and maintained.

Robustness is considered a separate factor altogether by many experienced protocol developers. RFC 1123 quotes the following robustness principle:

> Be liberal in what you accept, and conservative in what you send.

This principle of robustness is at the heart of all good software engineering, whether it relates to protocols, applications, or anything else. That's because good software engineering focuses on defining precisely what the user can and cannot do, and, by implication, what the program will and will not do. Moreover, as this robustness principle implies, good engineering includes a corresponding response to *every conceivable action* the user may take. Such all-inclusive measures protect equally against accidents and malicious acts, both of which are becoming ever more frequent occurrences on the Internet.

Since protocols normally have two basic components — actions that are local to your computer and associated actions on a remote computer — a highly conservative approach with respect to the expected behavior of the remote computer is considered prudent by experienced protocol implementers. Any such conservative measures must be explicitly programmed. The initiating computer's protocol must be written to expect certain operations and responses from the remote computer if the two computers are to interact and accomplish some task. But, at the same time, it also must make no assumptions about the specific capabilities of the remote computer. The initiating computer, for instance, has no way of knowing how busy the remote computer may be with other tasks, how well or ill equipped it may be to accomplish the desired task, or whether it has a relatively fast or slow network connection. It must make allowances for these highly variable conditions, and others just like them, if the expected interoperability is going to be achieved. Additionally, any number of bugs in a protocol's implementation or general problems in either computer's operation or performance might get in the way of the requested operation and might even escalate to the point of disrupting the network itself. Adhering as closely as possible to the fundamental, required elements of a protocol definition has proven itself to be the best way to avoid such problems.

The details on how the core, application-layer protocols function, which are presented in the following sections, will help to clarify why these considerations in defining and implementing protocols are so important. More importantly, they will clearly establish how the actions you routinely take on your computer, such as copying a file from a remote computer, are defined by and implemented in these protocols.

## The TELNET Protocol: Remote Access

The very first application-layer protocol was TELNET. Its purpose was simple, straightforward, and essential: to provide direct access to one computer from another across the network. It was important that TELNET be first because its primary function — gaining access to a remote computer — would be a required function of all host-to-host network operations. Before a file could be transferred from one computer to another, email could be exchanged, or a Web page could be loaded, the two computers first needed to start up a conversation and negotiate some form of remote access in order for the requested operation to be performed. The engineering problem that had to be addressed and resolved to make this basic interconnection possible, moreover, was as fundamental to TELNET as to all the protocols that would follow: how to compensate for or avoid altogether any architectural and operational differences between the two computers and find common ground on which to interact.

Like all application-layer protocols, TELNET exists in several distinctly different forms. Its first and simplest form is the protocol definition itself, as documented by its RFC. The RFC defines all the operations that a compliant implementation of TELNET must perform and the specific language constructs (e.g., *EHLO* signifies hello, *OK* signifies a successful reply, etc.) it must use in performing those operations. The RFC also defines additional, optional behavior, as explained earlier. TELNET also exists in the form of an application or program that functions as the user interface to TELNET. A user starts up such a TELNET program on his or her computer, and it is through this program that the user specifies the name of a remote computer to contact, logs on to that computer, and then enters commands and views responses. This form of TELNET is the one most likely to vary from computer to computer. It's here that programmers have the freedom to create anything from a very simple, command-line utility that assumes the user already knows how to use TELNET to a sophisticated application with a graphical user interface, drop-down menus, help screens, and settings for preferences. Such differences in TELNET implementations are possible because each program interprets the user's actions and translates or filters them into the prescribed language and operations specified in the protocol definition.

Most importantly (at least from the perspective of interoperability), TELNET exists as a computer process. This process is a type of operating system program — as opposed to a user program — running on a computer where TELNET is installed, and it consists of an implementation of the TELNET protocol written specifically for that computer's architecture. The TELNET process handles all the host-to-host interactions, such as initiating a connection with a remote host or answering an incoming request when a remote host is requesting access. It also handles all the user-to-host interactions on the local computer, interpreting all the user requests from the TELNET application and translating them into the appropriate host-to-host language defined in the protocol specification and implemented by the remote computer's TELNET process. In summary, the protocol definition specifies the rules and language necessary to establish interoperability between all different types of computers running TELNET; each computer's TELNET process implements the specification and handles all the network communication and user requests; and each computer's TELNET application provides the interface through which users can enter commands to access and use a remote computer.

TELNET functions by interconnecting the display and keyboard (known collectively as the terminal) on a local computer and the general operational environment (the operating system) on a remote computer. The result is that the remote computer considers the TELNET-connected user as if he or she were on a directly-connected or locally-wired terminal. In effect, the network takes the place of the wire and the protocol makes the user's terminal look like and operate like any other local and familiar terminal. This type of remote computer access was essential to establishing the time-sharing computing environment envisioned by the engineers of the ARPANET. TELNET made possible the sharing of expensive and isolated computer resources across a network by removing access restrictions imposed by the location of those resources and by enabling more people to use those resources at the same time.

TELNET, along with FTP, established the base functionality upon which many subsequent protocols would be built. Because most network protocols must negotiate remote access and manage the exchange of information and/or file content in the course of performing any more specific operations. Prior to the creation of browsers and the Web in the early 1990s, TELNET was used either directly or indirectly (through other programs that would transparently leverage TELNET's remote access capabilities) as one of the principal means of locating and accessing available resources on the Internet. One of the most common uses of TELNET was to access library card catalogs and other types of online databases.

A typical TELNET session looks something like the following:

1. You start up a TELNET application on your computer and enter the identification of the remote computer you want to access (e.g., telnet library.mycollege.edu). This initiates a TELNET session with the computer identified by its hostname, (i.e., library) and domain name (i.e., mycollege.edu).

2. DNS (one of the application-layer support protocols) silently looks up the IP address for library.mycollege.edu to locate the computer.

3. Your computer, functioning as the client, calls on TCP to establish a network connection with the library computer, functioning as the server (i.e., a networked computer housing information and/or services that others can access across a network).

4. The server starts up its own TELNET process in response to the requested network connection. Typically it responds by sending back its operating system identification and host name (e.g., SunOS Unix [library]) followed by a login prompt, which is displayed on your terminal by your computer's TELNET process.

5. You enter your login name, which is sent to the TELNET server.

6. The server responds with a password prompt, which is returned for display by your client.

7. You enter your password, which is sent to the TELNET server.

8. The server then evaluates your login and password against its database. If the information you have supplied is correct, the server starts a terminal session for your use. You can now access and use any and all of the facilities on the server that you are permitted to access through a local, directly-wired connection.

9. When you are finished with your session, you enter the appropriate command to end the session, (e.g., exit or logout), and the TELNET connection is terminated on both client and server.

What you see on your display in any TELNET session — the text you enter and the responses returned from the remote computer — represents only part of what TELNET is doing. The protocol is actually handling two distinct streams of data: one consists of your input and the TELNET server's replies that appear on your display; the other consists of messages sent back and forth between the TELNET processes of the client and the server. Think of these messages as a sort of shorthand taking place between the two computers as each computer's TELNET protocol implementation negotiates issues and communicates necessary information, the client process working on your behalf and the server process on behalf of the owners/operators of that host. For instance, the client has no way of knowing which optional protocol features are implemented by the server; and the same is true of the server with respect to the client. The client and server TELNET processes must, therefore, send messages back and forth to establish what each side can do, and what each wants, in order for them to successfully initiate and control your session. The exchange of messages is essentially a process of negotiation that lets each side signal what it can and cannot do. For instance, the client may send the message *WILL XXX* in order to signal it wants to start using option XXX, while the server may respond with the message *DO XXX* or *DON'T XXX* depending, respectively, on whether it does or does not implement that particular option, as shown in the sample client-server TELNET dialog later on in this section.

The TELNET protocol comprises the following four components, each of which is described below:

- The Network Virtual Terminal (NVT)

- Control commands

- Negotiable options

- Control functions

The Network Virtual Terminal (NVT) is the component most critical to implementing the interoperability of TELNET. That's because the NVT is responsible for absorbing any differences between the two interconnected computers. In the case of TELNET, these differences are localized to the type of terminal on the client computer, because all input and output is tied to the specific characteristics of that terminal. There exist today thousands of different types of computer displays, which vary considerably with respect to screen size, resolution, and many other features. The same is true of keyboards, which vary with respect to such basic particulars as the layout of their keys and the number and type of keys they contain, features often affected by the language and/or computer architecture they were designed for. Even at the time of the ARPANET's construction and the creation of TELNET, dozens of unique terminal types existed. Any two computers, therefore, could be expected to exhibit some number of differences, in varying degrees, between their respective terminal types.

Since TELNET interconnects the display and keyboard of a local, client computer and the operating system of a remote, server computer, the server's operating system, through its TELNET process, needs to recognize and display the letter 'G' when the 'G' key is pressed on any TELNET client, regardless of where that key might be positioned on the client's keyboard or how the client's operating system might represent that keystroke. Therefore, in order for TELNET to work consistently for all types of computers, and to avoid the unworkable solution of requiring each host to maintain a large, up-to-date database of all known types of terminals and their respective characteristics, something needed to be engineered into the TELNET protocol to compensate for or accommodate these expected variations.

That engineering took the form of the NVT, a virtual representation of a terminal that defined a lowest-common-denominator approach to keyboard and display characteristics that most if not all computers shared or could share. The NVT was to TELNET what the IMP-subnet was to the early ARPANET — an efficient, workable means to arrive at interoperability. The hosts on the ARPANET didn't need to be programmed with information about all the other types of hosts on the network in order to communicate with them. Instead, they only needed to be programmed as to how to interact with their local IMP, and the IMP-subnet would absorb any differences between the two computers. In much the same way, the NVT provided one, common definition of a terminal. Each local TELNET process, on the host and on the client, was programmed to interact with the NVT, which allowed the hosts to avoid a direct interaction when it came to interpreting a user's keystrokes or displaying information returning from the server. This, in turn, maximized and simplified interoperability.

Think of the NVT as a software implementation of a generic keyboard and display. Each computer's TELNET process uses its NVT to translate the specific constructs in its hardware and operating system into the generic representation defined in the protocol. The client and server can then communicate efficiently and in a highly predictable, reliable manner through this smaller, generic virtual terminal representation. In doing so, TELNET rises above any differences in architecture between the two computers. Both sides of a TELNET connection are, therefore, considered virtual terminals, or NVTs, since both client and server TELNET processes must translate user actions and server responses to and from their representation in the NVT.

The NVT serves another necessary function in addition to facilitating interoperability. The NVT is what enables the TELNET server to interpret your remote keystrokes as though they were local, and to transmit the data back in a way that allows it to be correctly displayed on your terminal. It takes the place of a physical terminal connected directly to the server, enabling the TELNET protocol's NVT to be viewed and treated no differently than any other type of physical terminal already known to the server's operating system, and thereby eliminating the need to make any changes to the operating system.

In addition to the NVT, the TELNET protocol documents the commands, options, and functions that define how the client and server TELNET processes shake hands, communicate, and control the flow of information. For example, commands include the WILL, WON'T, DO, DON'T operations for negotiating options; options include the ability to request and set the type of terminal to be used for the session; and functions provide control over simpler operations, such as erasing characters and lines, aborting output, and checking to see if the server is still connected. As shown in the example below, options include a further refinement that allows for lower-level suboptions to be negotiated, such as specifying a particular type of terminal the client may want to use.

Every command, option, and function is defined in the protocol with a unique name and an associated code. For example, the *erase character* function has the name EC and the code 247. Similar to DNS, in which numeric IP addresses are used to simplify and hasten computer-to-computer interactions while their associated host and domain names are used to assist human-to-computer interactions, the TELNET processes use the numeric codes to communicate back and forth while the names serve only to document those codes. The protocol also defines a special function designated as the *Interpret As Command* (IAC) character. As shown below, the IAC acts as a type of marker, signaling that what follows is a command sequence as opposed to data from the user.

The following sample TELNET dialog illustrates a simple exchange in which the client is requesting that the terminal type of the window be set to the value *VT100*, a popular, but not always implemented, terminal type configuration. The parentheses are used to identify the name associated with each code and are included for readability purposes only. TT stands for the *Terminal Type* option. SB and SE stand for *Subnegotiation Begin* and *Subnegotiation End*, respectively, and mark the beginning and end of suboption values. These markers are purposely generalized: by signaling the beginning and the end, they allow individual implementers of the protocol to customize optional features without disturbing the overall specification. They are able to place virtually any information in between these markers, and any compliant TELNET implementation will be able to respond that either it recognizes and supports that option, or it does not.

| TELNET Client-Server Dialog |
|---|
| Client 255(IAC),251(WILL),24(TT) |
| Server 255(IAC),253(DO),24(TT) |
| Server 255(IAC),250(SB),24(TT),1,255(IAC),240(SE) |
| Client 255(IAC),250(SB),24(TT),0,'V','T','1','0','0',255(IAC),240(SE) |

Line 1 shows the client asking the server if it allows terminal types to be specified (WILL TT). Line 2 shows the server replying that it does support this function (DO TT). Line 3 shows the server informing the client that a value for terminal type is required, indicated by the number 1 near the middle of the line. Line 4 shows the client responding that a terminal type setting is now being supplied, indicated by the number 0, followed by the specific terminal type, VT100, specified one character at a time.

The TELNET protocol is made powerful by its inherent simplicity. It handles the onerous task of accounting for all the computer hardware differences that exist by establishing, through the NVT, a virtual terminal representation through which a computer can be accessed and operated from any location. Meanwhile, it provides the communication structure, through control commands, negotiable options, and control functions, that allows these dissimilar and distant computers to talk to each other, negotiate differences, and establish a framework through which to exchange their data. If you step back in time for a moment and consider pre-network computers and the restrictive and isolated conditions of their use, you can better appreciate the magnitude of the advance represented by the creation, introduction, and widespread acceptance of the TELNET protocol.

## The FTP Protocol: File Transfer

The File Transfer Protocol (FTP) started out as a relatively simple, but powerful utility for copying files from one networked computer to another. Over the years, it evolved into a complex and sophisticated tool for such things as creating and sharing archived documents, distributing programs and program updates, and mirroring (copying and keeping synchronized) whole libraries of data at multiple Internet sites to provide faster and more localized file access.

Since FTP deals with files and file structures, its protocol specification is larger and more complicated than the one for TELNET. A wide variety of different file types exist, ranging from simple ASCII text files, such as a common README (.txt) file, to binary image files, such as a photograph's JPEG (.jpg) file, to large, compressed archive files, such as an application's ZIP (.zip) distribution file. FTP must negotiate a network connection to allow these files to be delivered or received, and it must adjust attributes of the file transfer to accommodate the different file types.

FTP was designed to meet the following four objectives:

- Promote file sharing.

- Encourage greater remote computer usage.

- Hide from users the differences in the way various hosts store files.

- Transfer data efficiently and reliably.

The following sections provide an overview of the architecture of FTP, discuss issues relating to the transfer of data and the exchange of files, and provide an example of what an FTP session looks like.

## The FTP Model

FTP uses two separate and distinct network connections — not unlike the different connections used in TELNET — to separate your interaction with a remote computer from the hidden computer-to-computer interaction. One connection is used for control purposes (i.e., exchanging commands and status information); it is referred to as the Protocol Interpreter (PI). The other connection is used for transferring the file data and for performing any associated operations on the file system (e.g., listing out files in a directory); it is referred to as the Data Transfer Process (DTP). Each connection uses a separate TCP session. As is the case with TELNET, both connections function according to a basic client-server model, with each assigned different tasks and responsibilities.

In general terms, an FTP process progresses as follows (the example in the section, "A Sample FTP Session," shows all the specifics). The client (i.e., your computer) initiates the control connection and sends FTP commands to the associated FTP server process. (Incidentally, this control connection follows the TELNET protocol. After all, the process is much the same: a handshake for a greeting, an exchange of information to establish who can do what, and so on.) The server listens for an incoming connection, completes the connection, and responds to the client's FTP commands over the control connection. The communication process consists of an alternating dialogue between the client and server. The FTP commands are interpreted by the server; they function first to configure the data connection and then to set the file transfer mode (i.e., sending or receiving data). The client then listens for an incoming server FTP process that signals the start of the file transfer. The server creates the data connection with the client's listening process, configures the file transfer and storage parameters, and transfers the data. The client is responsible for closing the control connection, although many servers now incorporate time-outs that check for a certain period of idle time and, if it is exceeded, close the connection.

## Data Transfer

The most demanding aspects of the FTP protocol are the lower-level issues surrounding data transfer. Because every operating system has its own unique mechanisms for storing and reading files, even something as simple as a plain text file (e.g., a README file) may require special handling across an FTP connection in order to display the same way on each computer. Something as basic as marking where one line ends and another begins is often handled differently from one operating system to the next. If you've ever retrieved a file from a remote computer and opened it to find one incredibly long line, you have encountered this particular problem.

The five principal data transfer components of FTP are listed here. In general, these features of the protocol deal with how the connection between the hosts is established and with specific issues related to the transmission of data.

■ **Data representation and storage**. This component deals with selecting the type and structure of the file to be transferred. Selecting the right data type is often critical to protecting the file data from being inadvertently damaged during the transmission. Selecting an inappropriate data type for a file may result in the file becoming corrupted and unusable. Although it is less commonly done, it is also possible to specify the file structure, which will affect both the transfer mode of the file and the local interpretation and storage of the file.

■ **Data connections**. This component deals with establishing the data connection between the local and remote ports and selecting the parameters for the file transfer.

■ **Data connection management**. This component deals with setting up a connection over non-default data ports and re-using open data connections.

■ **Transmission modes**. This component deals with selecting whether the data is transmitted with little or no processing (i.e., stream mode), is formatted into a series of data blocks with headers to facilitate retransmission in case of a problem (i.e., block mode), or is block-formatted and compressed for faster transmission (i.e., compressed mode).

■ **Error recovery and restart**. This component deals with enabling an interrupted file transmission to be retransmitted without resending the entire file.

## File Transfer

The client and server protocol interpreters control the file transfer process using a separate TCP connection. The FTP protocol defines all the commands that can be used by the protocol interpreters, how they are called, and what options, if any, they can take. These commands can be grouped into the following three areas:

■ **Access controls**. These commands deal with login and account information, moving around the server's filesystem, and logging out.

- **Transfer parameters**. These commands deal with the data transfer issues discussed above (i.e., specifying such things as a port or the data type, structure, or transmission mode of a file).

- **FTP service**. These commands deal with all the file-specific operations (e.g., specifying a filename to retrieve or send, listing files, making and removing directories, renaming files, and aborting a transfer).

An FTP command consists of a three- or four-letter string; examples are LIST, which lists out the files in the current directory, and CWD, which changes the working directory to a new directory. These commands are issued by the user's client FTP process, and each command is optionally followed by the space character (<SP>) and any command parameters. The TELNET end-of-line character (e.g., a carriage return) is used to signal the termination of the command information, which in turn causes the server's FTP process to interpret the command, perform any corresponding actions, and issue its reply. Since the control connection that handles the communication of information between the client and server FTP processes implements the TELNET protocol, the client's FTP commands and the replies from the server all follow the communication criteria set out in the TELNET protocol.

Every FTP command issued by the user receives at least one, and possibly several, replies from the server. An FTP reply consists of a three-digit numeric code, optionally followed by some explanatory text (e.g., 200 command okay). Each code represents a specific condition, and indicates the general success or failure of any command along with more detailed information. This exchange, or *conversation*, ensures that requests and corresponding actions are kept synchronized and that the user is always kept aware of the state of the server. The three-digit code greatly simplifies the process of programming automated FTP services. For example, FTP is frequently used by programs that run as part of an automated, nightly or weekly process to back up files from one computer to another across the network. Programmers can easily incorporate code into these programs that reads the numeric response codes, interprets their meaning, and then performs further operations based on the encountered condition. A response that the server is too busy might, for instance, cause the program to wait one hour and try again. The optional text that follows the number is intended to allow the conversation to be more easily reviewed and understood by a human audience, such as a programmer reviewing the log file from the previous night's file backup operation. This text may vary from computer to computer, since it is optional and has no bearing on interoperability. The numeric codes, however, must indicate the same general conditions to ensure interoperability.

Each position in the three-digit reply codes carries a different level of meaning, progressing from the most general condition represented in the first digit to the most specific information represented in the third. The first digit indicates that the response is good, bad, or incomplete; some programs may only need to check this digit in order to determine what to do next. If more information is needed, the program can check the second digit to determine, for instance, if the failure was due to an authentication error or a file system error. If the problem were a file system error, the third digit might indicate that the cause of the problem was a busy file system or there was insufficient storage space for the transfer to occur.

The following are some sample reply codes with their associated text:

| Sample FTP Replies | |
| --- | --- |
| Code | Description |
| 150 | File status okay; about to open data connection. |
| 200 | Command okay. |
| 220 | Service ready for new user. |
| 226 | Closing data connection; requested file action successful. |
| 230 | User logged in; proceed. |
| 250 | Requested file action okay; completed. |
| 331 | User name okay, need password. |
| 426 | Connection closed; transfer aborted. |
| 500 | Syntax error; command unrecognized. |
| 530 | Not logged in. |
| 552 | Requested file action aborted; exceeded storage allocation. |
| 553 | Requested action not taken; file name not allowed. |

## A Sample FTP Session

In the following FTP session, two different types of files — one plain text, or ASCII (the first standard for information exchanged, introduced in chapter 4) in format, and the other encoded, or binary in format — are retrieved from an anonymous (i.e., free service) FTP server. Anonymous FTP sites are common on the Internet, and have been since the early days. Grassroots Internet organizations use such sites, for example, to provide access to freely distributed software; and commercial organizations often use this type of FTP site to provide free and easy access to software updates or to trial versions of new software. As the name indicates, anonymous FTP sites are configured to grant anyone access to the files they contain, as opposed to requiring authentication through an existing account's login and password. Note that the following conventions are used below: user entries are printed in boldface text; plain text is used for replies from the server; square brackets enclose text that explains each event.

| **Sample FTP Session** |
|---|
| `ftp ftp.lobsterinfo.org`<br><br>[The FTP session is started by the user entering the local FTP program name, ftp, followed by the name of the FTP server to connect to. Just as with TELNET, any type of simple program or sophisticated application might be used to start up and run an FTP session. Currently, it's common for FTP sessions to be run within a browser window. |
| `Connected to ftp.lobsterinfo.org`<br>`220- The archives on ftp.lobsterinfo.org are`<br>`     an unsupported service of The Coon Cat`<br>`     Harbor Lobster Group.`<br>`     Use entirely at your own risk.`<br>`     ftp.lobsterinfo.org FTP server (Version`<br>`     wu-2.6.1(1) Mon Dec 18 13:09:40 EST) ready.`<br><br>[The server is located after DNS silently looks up the IP address of ftp.lobsterinfo.org. The server responds to the connection request with an indication that it is ready along with some general information about the site.] |
| `User (ftp.lobsterinfo.org:(none)): `**`anonymous`**<br><br>[The server transmits a User prompt, with "none" provided as the default account. The special user account, anonymous, is used; this is an optional feature defined in the protocol specification that servers can implement to allow anonymous users free access to their FTP site.] |
| `331 Guest login ok, send your complete e-mail`<br>`     address as password.`<br><br>[The server acknowledges that an anonymous, or guest, login is acceptable and indicates that the user's email address should be used for a password. This piece of Internet etiquette (frequently referred to as netiquette) is largely ignored; most severs don't even check for basic email address syntax.] |

| Sample FTP Session |
|---|
| Password: **crusty@crabs.com** |
| [The server prompts for a password, and crusty's email address is supplied as a response.] |
| 230- Welcome crusty@crabs.com ! It is<br>    Mon Jan 9 08:54:38 2004 local time.<br>230 Guest login ok, access restrictions apply. |
| [Login is granted and the server provides an ftp prompt (see below), allowing the user to enter commands.] |
| ftp> **ls** |
| [The ls command is entered at the ftp prompt to list the file contents of the directory. Servers vary in terms of what commands are made available. The help command can be used to display all available commands.] |
| 200 PORT command successful.<br>150 Opening ASCII mode data connection for<br>   file list.<br>crabfaqs<br>legalfaqs<br>lobsterfaqs<br>licenses<br>226 Transfer complete.<br>ftp: 256 bytes received in 1.4Seconds .18Kbytes/sec |
| [The directory contents are listed. Note that the listing itself is a type of file transfer, as the message indicates. That's because the FTP server is sending back data in the form of the file list, as opposed to the predefined reply codes and messages it has already sent. To send back this data, which is equivalent to it sending back the data contents of a file, requires that the FTP server process use the separate communications channel. The PORT command specifically acknowledges that the data channel has been opened.] |

| Sample FTP Session |
|---|
| `ftp>` **`cd lobsterfaqs`**<br><br>[The cd command is used to change directory to lobsterfaqs.] |
| `250- This is a mirror of the fyi.lobster.ed`<br>`250- archives.  Note: if you mirror this`<br>`250- directory, PLEASE exclude the file`<br>`250- allfaqs.gz.  It is a complete copy of`<br>`250- this directory and SHOULD NOT be copied`<br>`250- in addition to the other contents.`<br>`250 CWD command successful.`<br><br>[Each ftp directory can include a default message to be displayed automatically whenever someone enters it.  This directory displays such a message and then acknowledges that the command — CWD, or change working directory — was successful.] |
| `ftp>` **`get readme.txt`**<br><br>[Here the get command is used to retrieve the file named readme.txt.] |
| `200 PORT command successful.`<br>`150 Opening ASCII mode data connection`<br>`    for readme.txt (756 bytes)`<br>`226 Transfer complete.`<br>`ftp: 756 bytes received in 1.2Seconds .63Kbytes/sec`<br><br>[The PORT command again acknowledges the opening of the data channel.  Note that unlike the command interpreter's communication channel, which stays open, the data channel opens and closes as needed.  ASCII mode is used here to transfer the file, because it is the default and no other mode was requested.  File size and time required for the transfer are typical pieces of information supplied by the server.] |

| Sample FTP Session |
|---|
| ftp> **binary** |
| [The transfer mode is changed to binary, in preparation for transferring the second file, an application-specific file in the PostScript Document Format (PDF) file format. Binary transfer mode ensures that the file contents will remain unchanged by the FTP process. By comparison, ASCII mode might adjust line ending markers as appropriate for the local operating system, which is acceptable for a simple text file but would likely corrupt most other types of files.] |
| 200 Type set to I. <br><br> [The server acknowledges the transfer mode change, changing the file type to I for Image. (The earliest binary files were images, hence the protocol's use of I and Image rather than the more generic B and Binary.)] |
| ftp> **get lobsterfaq1.pdf** |
| 200 PORT command successful. <br> 150 Opening BINARY mode data connection for <br>     lobsterfaq.pdf (128 bytes) <br> 226 Transfer complete. <br> ftp: 128 bytes received in 1.9Seconds .64Kbytes/sec <br><br> [The only difference here is the acknowledgment that BINARY mode is being used for the file transfer.] |
| ftp> **quit** <br><br> [The session is terminated with the quit command.] |
| 221- You have transferred 1984 bytes in 2 files. <br> 221- Thank you for using the FTP service on <br>     ftp.lobsterinfo.org. <br> 221 Goodbye. <br><br> [The server supplies general information about the session before closing the connection.] |

The session displayed above makes it clear that FTP is fundamentally a two-way conversation between client and server. The exchange of information is continuous, with each action by the client generating one or more responses from the server. This simple, orderly approach is not only responsible for the popularity and efficiency of FTP, but it also provides a clear, concise, and consistent framework that programmers can leverage in their creation of utilities for automated and/or repeated file transfer operations. The three-digit reply codes supply the necessary status information for these automated programs to function. These codes allow a program to read and respond to conditions as the FTP process progresses from initial connection to locating, accessing, and transferring a file. They also contribute to the accuracy and usefulness of log files that are often incorporated into these programs to record their events for later review. The codes serve to confirm successful file transfers and to document specific error conditions that may need to be examined and resolved.

You can also see in the session above how the commands entered by the user differ from the three- or four-letter command strings described earlier. For instance, the user enters *ls* to list the files in the current directory, while the protocol defines this command as LIST; and the user enters *cd* to change directories, while the protocol defines this command as CWD. This separation between user command names and protocol command names is precisely the same as the one described earlier with respect to TELNET. The implementation of FTP, like that for TELNET, is divided into a user component and an operating system process. The user component, whether a simple, command-line utility or a large-scale graphical application, handles all the user interactions. This allows software vendors to customize their product implementation of the protocol for any audience (e.g., experienced or novice, French or English speaking). The user software maps more easily understood or familiar command names to the underlying, protocol-specified three- or four-letter commands, which are then used in the computer-to-computer interactions that constitute the FTP client-server communication stream.

The power, versatility, and simplicity of design found in the FTP protocol specification has contributed to the success of the Internet in innumerable ways. The fact that FTP continues to be a driving force behind the sharing of information on the Internet is the surest indicator of the soundness and robustness of its engineering. It's unlikely that the Web would be what it is today had it not been for the simple, elegant, and refined engineering of FTP.

## The SMTP Protocol: Electronic Mail

The Simple Mail Transfer Protocol (SMTP) conveys the single greatest reason for email's widespread, unrivaled success right in its name: simple is always better when you want to maximize interoperability and encourage general use. Since electronic mail, or email, has done more to bring people to the Internet than any other single feature, email's history and the specifics of the SMTP protocol are discussed in detail in the following chapter. But a few points are better made here.

Unlike the protocols described above, email had an unplanned beginning. At the time the ARPANET was being built, many computers already included some type of rudimentary email system. Those systems, however, were limited to sending and receiving messages to and from accounts on the same computer. But when the ARPANET came to life in the early 1970s, with it came the ability to ship files from one computer to another through FTP. It was not long before many people began experimenting with how this networked file sharing could be applied to the exchange of personal messages or general correspondence to and from accounts on different computers attached to the network.

Just as the core functionality of FTP was initially packaged into the TELNET protocol, email became part of the FTP protocol in 1973 and remained there until the early 1980s when the separate SMTP protocol was created. FTP was already designed to handle most of the work that would be required to create network-based email. Moreover, FTP was designed for use by automated programs, which is precisely what was needed to manage the exchange of mail between computers. If you strip away the user's interaction with email (i.e., composing and reading messages), you are left with the networking component (i.e., shipping a message file from the sender's computer to the designated computer address of the recipient). An email program only needed to interpret an email address (i.e., distinguish the user or mailbox name from the destination computer's address), call on FTP to open a connection to that computer, change directory to one that held the mailbox of the recipient, transfer the message file, and quit the connection. This was how networked email began. Fundamentally, not a lot has changed since, as chapter 6 will describe.

## The DNS Protocol: Domain Name Translation

The Domain Name System (DNS) is an integral and necessary support function of the Internet. The system as a whole simplifies both network management and the routing of data across the Internet by doing the following:

- Defining how host and network names are mapped to IP addresses.

- Establishing a hierarchical system for locating IP addresses.

- Providing a system whereby networks are arranged into a hierarchy and subdivided.

DNS includes the definition of a host-resident protocol that implements the translation of host and network names into IP addresses, which in turn allows the location of Internet hosts to be identified and data to be delivered to the correct address.

Because the Internet comprises millions of computers, a simple and efficient system for naming and locating hosts is essential to its operation. Given that the Internet is decentralized (i.e., without any single or central point of operation) and distributed (i.e., the individual networks that constitute the Internet are controlled and administered locally, anywhere in the world they may exist), DNS was designed to be hierarchical in nature. The hierarchical structure allows individual network administration to remain local, while it provides the necessary controls at higher levels to maintain consistency across the network as a whole (i.e., host/domain names and IP addresses must be unique across the entire network). DNS was engineered to meet these needs.

The reason DNS is a core Internet protocol is that every host on the Internet must implement some basic DNS functionality in order to interact with other computers on the Internet. For instance, every host must minimally implement something called a DNS resolver and use this resolver to convert host names to IP addresses. DNS resolvers are used to query domain name servers (i.e., servers that store a portion of the database that contains the host name and IP address associations). Due to the number of database entries and the frequency with which these associations change, the design of this database is distributed and hierarchical in structure. Each domain name server holds only a very small portion of the database, but is configured to query other domain name servers when it cannot provide an authoritative response to a query. In this way a single query from your computer's DNS resolver can trigger one or more queries through a chain of domain name servers scattered across the Internet until an authoritative response is found and the needed IP address is returned.

The simplest instance of the use of DNS is also probably the most common. When you connect to your Internet service provider (ISP), they create an IP address and computer name for your host on the Internet. This is called dynamic IP allocation, since this process is repeated each time you connect to your ISP. When your ISP allocates your name and address, it updates its domain name server with this information. It is this action that makes your computer into a connected and reachable host on the Internet. What also typically happens when you connect to your ISP is that they will dynamically configure the location of a local domain name server for your computer's DNS resolver to use in performing DNS queries. Your computer minimally needs the IP address of one domain name server in order for it to connect to other hosts on the Internet. Otherwise, it would have no means of converting the host name found, for instance, in a Web site's hyperlink or URL (e.g., http://lobsterinfo.org) into an IP address. Every time your computer needs to transmit some data over the Internet (e.g., to send an email, start an FTP session, or request a Web page), the DNS resolver is called into action to look up a host name and return an IP address. The operations of the DNS resolver and the operations of a domain name server are defined by the DNS protocol.

Because host names must be resolved into IP addresses before data can start traversing the Internet, DNS represents one of the most frequently used services of the Internet. Only on rare occasions do most Internet users even catch a glimpse of its existence, and only then when some lookup fails and an error message is displayed that a particular host name could not be found. One reason DNS is so fast and dependable is that the protocol defines a caching mechanism (i.e., a temporary store of information) that considerably reduces overhead on the network. When domain servers query up the hierarchy of domain servers to search for an IP address, they can be configured to retain the information they obtain for some limited period of time so that additional requests for the same information can be answered locally instead of requiring another long-distance search. Because of the highly dynamic nature of the Internet, however, such caching must be temporary. To understand why, think of your own computer coming and going on the Internet. Your IP address and name may change each time you reconnect. If those cached entries were allowed to remain on any number of remote DNS servers for any length of time, a lot of stale and inaccurate DNS entries would accumulate and this would result in a lot of misdirected traffic.

Another interesting aspect of DNS is that it is normally configured to use UDP (the smaller, simpler version of TCP) rather than TCP. Because UDP, unlike TCP, was designed not to be concerned with transmission errors or with providing any reliability functions to compensate for the inherent unreliability of IP, it offers a lower overhead and a smaller packet size than TCP. This makes possible a faster response time and a generally lower network overhead. DNS's use of UDP is critical to the overall functionality and efficiency of the Internet. Anything and everything that reduces the demands of DNS on the network helps to lessen its impact on the Internet's overall performance.

Precisely because DNS is such an active service, the protocol is very clear on imposing restrictions on both the DNS resolvers and the domain name servers regarding the retransmission of queries. Limitations on queries are essential to ensure that the Internet's bandwidth is not wasted. Accordingly, the protocol specifies that queries that receive no response after several attempts must return an error, rather than continue to resend the request; and that failures are to be cached temporarily to avoid repeating the same failing query again and again. Various algorithms are used to determine retransmission rates, in order to provide a best effort at obtaining the requested information while keeping DNS's impact on network performance to a minimum.

DNS may be the most highly active Internet service that you virtually never see. Its protocol provides the means to quickly and efficiently locate any host on the Internet anywhere in the world, even as any number of those hosts come and go on and off the Internet every minute of every day.

## Host Initialization

Host initialization refers to the process of using the network as part of starting up and configuring a computer. Unlike most computers, which consider the network an adjunct service and are designed to operate equally well off the network as on, host initialization is used by computers that are purposefully built with limited or no capacity to store permanent information (e.g., they have no disk drive). Accordingly, diskless workstations and other similarly designed computers consider the network an integral component of their architecture and, as such, necessary for their operation. The protocols created to meet the needs of host initialization define how these computers can send out messages onto the network and discover the information they need to configure themselves and become fully operational.

The introduction of the personal computer in the early 1980s set the standard for a computer-use paradigm and a general computer architecture that has remained remarkably unchanged for some time, despite considerable advances in computer technology, the commercialization of the Internet, and the widespread adoption of networking. Computers from twenty years ago and computers today contain much the same basic hardware, software, and operating system components: an output device for displaying information (e.g., a monitor), input devices for entering information (e.g., a keyboard and mouse), a central processing unit (CPU) for handling instructions and performing calculations, one or more memory modules for storing temporary information, one or more disk drives for storing permanent information, locally installed software for creating and viewing files, and so on. All of these components are repeated ad nauseum from machine to machine; yet, despite all this redundancy, our use of these computers is normally tied to one or two specific machines that contain our individual account information, files, and personal settings. These redundancies stand in stark contrast to the resource-sharing design goals of the ARPANET and early Internet and the underlying objective of leveraging the network to introduce all sorts of new efficiencies and freedoms. Work has been done, however, in an effort to address this situation, with promising results and yet limited impact.

Diskless workstations, for example, represent a different paradigm relating to both computer use and computer architecture — one that views the network as *the* principal resource. Diskless workstations first appeared in the 1980s as an alternative to personal computers and a network-based, evolutionary improvement over the older dumb terminals (i.e., simple character-based or graphical, bit-mapped displays with keyboards wired directly into a computer center's servers). They formed part of a client-server networked architecture that leveraged the network and the computer center's servers to perform all the same operations that other types of more expensive workstations, and personal computers, could perform. Moreover, diskless workstations were simpler devices, typically less expensive to purchase, always less costly to maintain, and less prone to failure. Furthermore, they took advantage of the larger and more powerful resources available through one or more networked servers, enabling them to out-perform personal computers; and, because all information was stored on the servers as opposed to the individual workstations, a diskless workstation could be used by any individual without requiring the type of customization demanded by personal computers and frequently required by other types of workstations.

The argument behind choosing a diskless workstation over a personal computer was that, since it was now possible to network computers and share files and other resources, it was no longer necessary to incur the added expense of having each computer repeat the functions of all the other computers on the network. Instead, it made sense to dedicate a few computers to function as servers and collectively store users' files and provide some or all of the most commonly used applications in one or more centralized locations. Why install, for instance, a word processing or drawing application on dozens or hundreds of individual computers when you could instead install it on a single server, and provide access to the application from any computer on the network? The other computers (i.e., the clients) could then be simpler, less expensive, and far easier to configure and maintain. Such a configuration would have the added benefit of reducing the overhead associated with backing up files and installing new or updated applications.

Networked appliances are in many respects the latest incarnation of diskless workstations, but designed with the Internet in mind. Like diskless workstations, they rely utterly on the network, both to start up and to operate. Host initialization protocols specify how these types of diskless computers, and other computers with limited storage facilities, can use the network like an extended disk drive to acquire information from network-accessible resources that it needs to initialize and become ready for use. For diskless hosts, these protocols require that both of the following processes be performed. Hosts that have disks but that function primarily as networked clients may only need to perform the first process.

1. **Configure the IP layer**. In this process, minimal information is configured on the client so that it can find its host and boot server. These servers (which can be the same computer) then transmit the networking information the client needs to establish itself on the network.

2. **Load the host system code**. In this process, once the network connection is configured, a file transfer protocol is used (e.g., FTP) to copy operating system code from the server so the client can finish initializing.

Various protocols specify exactly how these processes can be performed. Adherence to these protocols, like elsewhere, ensures interoperability and allows manufacturers of these devices to engineer and build their products according to established standards. For example, the following protocols are commonly used to support dynamic configuration across a network:

- **ICMP Information Request/Reply Messages**. This protocol enables a host to query the local network for its identification.

- **Reverse Address Resolution Protocol (RARP)**. This protocol provides a broadcast medium that allows a host to find its IP address by providing its link-layer address.

- **ICMP Address Mask Request/Reply Messages**. This protocol provides a message-based service that allows a host to uncover the address mask (a unique identifier) for a specific network interface.

- **BOOTP Protocol**. This protocol allows a host to broadcast a request via UDP to find the local BOOTP server, which in turn allows it to find its IP address, the name of an appropriate boot file, and additional network information.

It's likely that this protocol area will grow considerably in the near future as this type of client-server architecture becomes more prominent, first in business environments, and eventually in the home. The network itself, in the form of the Internet, workplace intranets, and even modest home networks, will reduce the complexity, redundancy, and expense of computer equipment. It will achieve this by facilitating a far more expansive and fundamental sharing of computer resources, just as it has already facilitated a sharing of information. Protocols, like those mentioned above, will help considerably to make such changes possible and to make the transition easy if not effortless for most individuals.

## Remote Management

Remote management refers to the process of using the network to monitor and manage computers, other networked devices, such as routers, and the network itself. Remote management protocols installed on network devices define how messages are communicated over the network to allow the devices to be monitored and controlled from across the network. These protocols also define the types of operations that can be performed remotely and how data is collected for accounting purposes, analyzing the network's performance, locating and resolving problems on the network, and other purposes.

Accompanying the proliferation of computers and the escalating reliance on their use, especially in the workplace, is the issue of how to manage networked computers and other networked devices. Personal computers in particular are costly and time consuming to configure, monitor, and maintain. They require that applications be installed locally, which results in both the application files and the files created by the user of the computer needing to be stored on the local disk drive. This, in turn, presents additional cost and maintenance issues. For example, it requires backing up data from each computer to protect against data loss in the event that the local disk drive fails. As the number of personal computers increase in any environment, large or small, the time, effort, and costs related to their management, and to the management of the network, also increase. Few options, other than the network itself and remote management, are capable of providing large-scale remediation of these issues or any economies of scale.

The network and remote management protocols together provide a bold, new, effective means to monitor and manage both computers and the network. Just as networked appliances and diskless clients aim to reduce operating costs and overhead, the remote management of computers and computer networks is another growing area that promises to increase the use and value of the network and simplify the administration and monitoring of the ever increasing number of networked devices. Among the protocols defined to implement the management of Internet devices, including hosts, are the following:

- Simple Network Management Protocol (SNMP)

- Common Management Information Protocol Over TCP (CMOT)

These protocols define messages that are used in the management of a network. Managing a network involves the constant monitoring of devices and traffic, along with quick problem resolution, to keep the network up and running smoothly. Messages collected from the devices on the network constitute the data used by programs to monitor the network. Messages are also used to perform specific operations on networked devices, such as shutting down or restarting a router that may be behaving erratically. As a network grows, the capability to monitor and manage individual computers, and groups of computers, from a remote location becomes more and more important. Even the IMPs of the ARPANET were designed with self-diagnostic and monitoring functions in order to protect the overall network from localized failures. SNMP and CMOT allow thousands of network devices to be analyzed and monitored from a single location; the information they collect facilitates problem resolution and helps prevent small problems from becoming large ones.

In general terms, three components are needed for remote management:

- **Network Device**. This managed object is typically a host or gateway that is part of a network and requires monitoring and management. It must include a management agent in order to collect information and pass it along to the manager.

- **Agent**. This program, which is installed on each network device, collects and maintains information about the local network device, stores and retrieves management information from a configured management information base (MIB), and moderates communication between the device and the manager.

- **Manager**. This program operates on the system doing the monitoring and serves to query network device agents, collect and set data, and acknowledge events.

Typical uses for remote management are:

- **Fault management**. The process of discovering, isolating, and correcting aberrant behavior.

- **Accounting management**. The process of assessing cost according to network use.

- **Configuration and name management**. The process of performing initialization and shutdown operations.

- **Performance management**. The process of analyzing network behavior and performance.

- **Security management**. The process of storing and distributing passwords and encryption keys.

The various protocols for initializing and running hosts across a network, as well as for managing hosts across a network, illustrate the growing importance of the network's role as facilitator. They also point to the network's vast potential for reducing both operating costs and the complexity and extensive duplication of hardware and services commonplace in our current use of computers. These protocols, and others currently in development, will lead the way to the future, a future in which the network itself will provide for all of our data and communication needs, and computing devices will be greatly simplified, consisting primarily of displays and keyboards, or other control mechanisms, and, of course, a connection to the network.

# Email: The Unforeseen Catalyst

## A Network of Computers Becomes a Network of People

It is hard to imagine the Internet without electronic mail, or email. Email has been around longer than the Internet and longer than networking itself, and was adapted to work with networking only a few years after the ARPANET started transmitting its first packets of data. Since that time, email has been the most commonly used network service, across all age groups, races, economic groups, geographic regions, and types of users. There has been a great deal of study and speculation regarding the reasons for the overwhelming and widespread popularity of email. Its simplicity, which is considered by many its most compelling feature, may be the most important reason behind its success; or perhaps it's the inherent and enduring capacity of email to bring people closer together, no matter the distance that separates them or the ever increasing busyness of their lives, that accounts for email's strong, universal appeal. What is certain is that email is more responsible than any other Internet service for enticing people onto the Internet and keeping them online; and this is not likely to change anytime soon.

The Internet is first and foremost a communications network. But it was engineered to connect computers, not people. Its goal was to provide direct and shared access to scarce and expensive computer equipment with the intention of facilitating the sharing of computational power and resources. But email unexpectedly introduced a human element onto the network and, in doing so, turned the Internet into a communications network of a different sort — one that connected people. This, more than any other single event in the history of the Internet, transformed the Internet from a technological achievement, an information-age tool for the pursuit of scientific, mathematical, and military objectives, into a force for societal change. Email turned the Internet into something personal and something anyone could use, which in small, tangible ways as well as subtle, far-reaching ones, changed the way we live, communicate, and interact with others.

The history of email is characteristic of the history of innovation on the Internet. Like so many other Internet creations — the Web is another good example — email seemed to come from nowhere, become immensely popular in a remarkably short period of time, and, most importantly, alter the way many people regarded the Internet, its potential, and its future. Email was never considered one of the objectives in the building of the first networks in general, or the engineering of the ARPANET in particular, because email had no place in the time-sharing model of computer use. Time-sharing viewed existing computer resources from the perspective of faster, more immediate access to raw computational power. The network would supply researchers with more economical and more egalitarian access to those resources. It was a means to an end. Unlike today, there was little or no recognition of the network as a resource in and of itself. Nor was email part of the military's objectives in their funding of ARPA, the ARPANET, and the early Internet. Their focus was exclusively on creating distributed and survivable command and control functions. Instead, email turned out to be a spontaneous and organic by-product of the early ARPANET, a facility of the network itself waiting to be discovered and exploited.

There is another, more fundamental reason why email was never included in the planning of the ARPANET. At the time, computers were scarce. Any type of personal computer use was a decade or more away, and no one at the time envisioned that such a future might be so near. If nothing else, there were far too many technological and economic hurdles standing in the way of the mass production of small and affordable personal computing devices of any sort. So, even though the Internet is a tribute to the fact that the ARPANET engineers were thinking big in terms of their engineering, their plans for the network were appropriately modest in scope and focused entirely on resource sharing. At the end of 1969, there were only the initial four nodes, while two years later there were just fifteen nodes that connected twenty-three host computers on the network. There were not enough computers, or people who routinely used computers, to make interpersonal communication a factor. Accordingly, the goals of the ARPANET engineers were to link up computers and to develop the infrastructure and tools needed to establish computer-to-computer communication. Human communication across the network developed simply as a natural, albeit unexpected, by-product.

It was in 1971, when there were just twenty-three host computers on the network, that Ray Tomlinson from BBN invented the first network-based email program. His creation was the first of many network-driven utilities that were largely cobbled together from existing components. In this instance, Tomlinson combined an intra-computer messaging program (called SNDMSG) with an experimental file transfer program (called CPYNET) to enable the first mail messages to travel from computer to computer across the network. Once it had been introduced, people immediately recognized that this new capability was something far more than the sum of its parts. The network had taken on the quality of a catalyst; its existence gave new dimension and purpose to existing functions and services, even to something as simple as an electronic message. A few short years later, email programs were showing up all over the network.

In 1973, the first large step towards establishing the interoperability of email was taken when the FTP protocol — as defined in RFC 454 — formally incorporated email into its specification by defining both an independent mail command and the special characteristics of a mail file. This first formal email specification, which used FTP to handle all network operations, worked simply, quickly, and efficiently. It worked so well that it took another eight years before a separate, highly standardized mail protocol was created; it was called the Simple Mail Transfer Protocol (SMTP).

Email's early and eager acceptance on the ARPANET did not just impact the way in which the network was being used or, more generally, how the network was regarded. It affected how people communicated and interacted, much like the invention of the telephone had done roughly a hundred years earlier.

Communicating in email quickly acquired its own particular, and even peculiar, form of etiquette and linguistic elements. In 1978, J. C. R. Licklider observed the following while contemplating the differences between email and other forms of communication:

> It soon became obvious that the ARPANET was becoming a human-communication medium with very important advantages over normal U.S. mail and over telephone calls. One of the advantages of the message systems over letter mail was that, in an ARPANET message, one could write tersely and type imperfectly, even to an older person in a superior position and even to a person one did not know very well, and the recipient took no offense. The formality and perfection that most people expect in a typed letter did not become associated with network messages, probably because the network was so much faster, so much more like the telephone.

> Among the advantages of the network message services over the telephone were the fact that one could proceed immediately to the point without having to engage in small talk first, that the message services produced a preservable record, and that the sender and receiver did not have to be available at the same time.[1]

Email, as a new and different communication medium, had quickly developed its own unique qualities and uses. But because it had not been a planned feature of the network, the spread and development of email was largely without any bounds or control. Several RFCs from the mid 1970s document some of the efforts made by the ARPANET and Internet engineers in trying to curtail the unrestrained expansion of email and exert some measure of control over an all important network service that was exhibiting a life of its own. RFC 808 summarizes a meeting held at BBN in early 1979 that focused on the existing computer mail services and the issues and concerns related to the future growth of email. The following brief excerpt shows just how out of control the situation had become:

Dave Farber gave a bit of history of mail systems, listing the names of all the systems anybody had ever heard of (see appendix A). It was noted that most of the mail systems were not formal projects (in the sense of explicitly sponsored research), but things that "just happened."[2]

In all, some thirty-six different mail systems operating on more than a dozen different machine types were listed in the appendix mentioned by Farber. The list illustrates how far and wide email had spread in those early years, even though the network was still relatively small and computer access was largely limited to research facilities and universities.

Not until the late 1980s did email and the commercial interests of business start to intersect. In 1988, the first sanctioned commercial email started being transmitted across the Internet when MCI Mail was allowed to connect to the NSFNET through the Corporation for the National Research Initiative (CNRI). The connection was supposed to be temporary and experimental. This was, after all, several years before the commercialization and privatization of the Internet. But the experiment's immediate and unqualified success meant that commercial email was here to stay; there was no turning back. MCI's email service was quickly followed in 1989 by Compuserve getting approval to connect its mail system to the NSFNET. In 1993, America Online and others followed MCI's and Compuserve's lead and began connecting their proprietary email systems to the Internet. It was then it became clear that global Internet email standards would be in everyone's best interests, and the pursuit of such standards quickly acquired more attention and more relevance.

Since the time of the first standards work on email, when email became part of the 1973 FTP RFC, a lot of standards documents have been written that directly or indirectly relate to email. These documents cover such things as the transmission of email, message formats, authentication mechanisms, how to combat junk mail (i.e., spam), content encoding, multimedia files, encryption algorithms, electronic signature policies, internationalization, and even email's interoperability with facsimile devices and calendar systems. Fortunately, the core functionality of email can be divided into the following four areas:

- **Email transportation**. This area covers how mail messages find their way through the Internet to their destination. This process is defined by the SMTP protocol.

- **Receiving and sending email**. This area covers how email clients, like those on a home PC, communicate with email servers. Most clients use either the POP3 or IMAP4 protocols for this.

- **Email message format and encoding**. This area covers what your email looks like, from the computer's perspective.

- **Mail extensions, or MIME**. This area covers how email allows you to attach and incorporate multimedia files (and all types of other files and different forms of content) into your messages.

These subjects, starting with SMTP, are described in the following sections. Following that, some commonly misunderstood email uses, abuses, and concerns are presented, including: mail forwarding, unsolicited email, viruses carried in email, email privacy, and email security.

# Email Transportation

The Simple Mail Transfer Protocol (SMTP) defines how mail servers communicate with each other in order to transport mail messages across the Internet. The protocol owes much of its general behavior to FTP, which is not surprising given that one of the main reasons for creating SMTP was to remove all mail functions from the FTP specification and continue email's development as a separate Internet service. The functionality of SMTP, therefore, will seem familiar. Just like FTP and TELNET, SMTP operates according to a dual-component, client-server model that establishes and maintains a conversation between two computers across a network; one component handles the transfer of data — a mail message in the case of SMTP — and the other handles the negotiation of control commands and functions.

The transfer of mail via SMTP is considered an *end-to-end delivery* service, because it requires the sender's SMTP email server to connect directly to the recipient's SMTP server. This type of service differs from the *store-and-forward* delivery service used in other email systems, like those using UUCP (or, for that matter, other large-scale message delivery systems, like the telegraph or postal network), in which messages are passed through a series of intermediate hosts before arriving at the recipient's email server. SMTP's direct delivery approach is largely responsible for ensuring that the service meets its primary goal: "to transfer mail reliably and efficiently."[3] Email can, however, also be delivered through intermediate relays or gateways that accept the incoming mail and take responsibility for its final delivery. This situation typically exists when email has to pass through dissimilar networks, or when a firewall or gateway mail server has been configured to collect all incoming email for a domain and then forward it through the connected intranet, as is commonly done in corporations and other organizations that have large computing environments. The SMTP specification defines how such indirect delivery mechanisms should function in order to maintain the necessary reliability of the overall system.

Even though a user does not interact directly with SMTP (as is done with FTP and TELNET), SMTP is still an application-layer protocol. As such, it relies on a protocol in the transport-layer (the layer below the application layer in the network stack) to provide a reliable means of transmitting its data. Typically, TCP performs this service on the Internet, but any similar service could be used.

The remainder of this section first describes the sequence of events that occur once your mail message reaches the mail gateway attached to your local network; it then presents a sample SMTP session that shows precisely what the SMTP client-server conversation looks like. Note that a mail gateway is a type of computer configured for your business location, or for the network operated by your ISP, to handle the incoming and outgoing flow of mail messages. How a message gets from your computer to the mail gateway is described in the next section. Note also that the terms client and server used below are a little misleading in the case of email delivery. Think of the client as the initiator or originator of the process and the server as the destination mail gateway for the message.

When an SMTP client (the local mail gateway) receives an outgoing email message from you, the message goes into a general queue of messages awaiting delivery. When your message's turn is reached, the SMTP client starts up a two-way communication channel with an SMTP server. The destination (or server) SMTP host must first be located by means of a DNS lookup based on the recipient's email address contained in the message. This lookup either returns the IP address of the recipient's local mail server as contained in the address, or it returns the IP address of another computer altogether, which is often the mail server or gateway that has been identified in DNS as the email collection point for the recipient's domain. DNS makes this distinction by providing tokens that network operators can use when they list their network's host names and IP addresses on their DNS server. This simple mechanism allows one mail server to collect and distribute mail for any number of local email accounts, whether or not those accounts use the same domain name in their email addresses. For example, email sent to the addresses *scotty@maine.siamese.com* and *dexter@vermont.siamese.com* may both be delivered to the machine *mail.siamese.com*, provided it is identified in the company's DNS listings as the mail gateway for the siamese.com domain.

Once the communication channel is open and the handshaking between the two SMTP hosts has been completed, the SMTP client initiates the actual mail transaction. The objective of the mail transaction is simply to transport a mail object. As is described in more detail in a later section, this mail object has two distinct parts: the envelope (which consists of the originator's address, one or more recipient addresses, and control information that is used to negotiate optional features for the data transmission) and the content (which contains the message headers and the body of the message).

The transaction itself consists of commands and replies that are strikingly similar in structure to the protocol interpreter dialogue transmitted during an FTP session. Just as with FTP, each command necessitates some reply. The reply may indicate that more information is expected or that some error has occurred, or it may denote successful completion. First, the client sends information about the originator and the destination (i.e., your email address and the address(es) of the intended recipient(s)), then it sends the message content, including the message header and any other structural information.

If the same message is sent to more than one person with an address at the same destination domain, only one copy of the message will be transmitted to that domain for all the local recipients. If the same message is going to more than one domain, however, the entire process must be repeated for each domain. Once the message has been transmitted and acknowledged, the client may request that the channel be closed or it may start up another mail transaction, if it has more mail in its queue for that domain. The client may also use the open channel for additional services, such as verifying the receipt of specific emails or retrieving subscription mailing list addresses.

Each part of the process outlined above is described in a little more technical detail below:

1. **Session initiation**. The client opens a connection to the server SMTP and awaits a reply. If all is well, the server returns a 220 code to accept the connection. Typically, the server also sends back its name and sometimes a greeting as part of its acknowledgment. Just as with FTP, the text that accompanies the three-digit reply code is strictly for human consumption and is particularly valuable when examining log files for problems.

2. **Client initiation**. The client now sends the EHLO command (a form of *hello*) followed by the client's name and, optionally, by information that conveys to the server any additional features supported by the client's implementation of SMTP. If all is well, the server responds with 250 OK to confirm that it is ready to accept the mail. It, too, may send along a list of features its implementation supports. The SMTP client and server are now in the *initial* state, which means that no transaction is active and that the state tables and buffers (i.e., the areas that cache, or temporarily store, the data for the transaction) have been emptied.

3. **Mail transaction, the MAIL command**. The client starts the transaction by sending the MAIL command, which is defined as follows:

```
MAIL FROM:<reverse-path> [ <mail-parameters> ]
```

The *reverse-path* contains the sender's name (or mailbox) and will be used by the server if errors need to be reported or receipt verification has been requested. Optional *mail parameters* are also specified here and relate to whatever features were negotiated during the initiation process. A response of 250 OK confirms that the transaction can continue. If there is a problem, the server must communicate to the client whether the problem is temporary (in which case it should try again later) or permanent (in which case it should return an error condition to the sender and not retry the transaction).

4. **Mail transaction, the RCPT command**. The client now sends the RCPT (i.e., recipient) command, which is defined as follows:

```
RCPT TO:<forward-path> [ <rcpt-parameters> ]
```

This command uses *forward-path* to identify the recipient of the message, normally in the form of a mailbox and domain name, such as *dexter@vermont.siamese.com*. If the recipient is known, the server responds with 250 OK; otherwise, it returns a code 550 and, typically, some text that states that the user is unknown. If there is more than one recipient of the message at this server's domain, this step — checking for and acknowledging a recipient — is repeated for each one.

5. **Mail transaction, the DATA command** Now it's time for the client to start sending the message. It begins by sending the following simple control line:

```
DATA
```

The server accepts transmission by returning a code 354 and then treats all subsequently transmitted lines as the message text. An end-of-mail indicator, consisting of a dot '.' on a line by itself, marks the end of the message's content. This indicator also implicitly acts to confirm the mail transaction, which in turn signals the SMTP server to process the data for the stored list of recipients. The server returns a 250 OK reply when it has finished accepting the data. In processing the message, whether for local delivery or for relaying to another server, the server inserts a trace record containing host information identifying the client and server along with the date and time the message was received. Such trace records take the form of *Received:* header lines, as shown below, and are typically used for debugging mail transfer faults. In addition, if this is the final delivery of the message, (i.e., no further relaying or forwarding is being done), the SMTP server adds a single *Return-Path:* header line at the beginning of the mail data. Only one such line should exist in any mail message; it is used to preserve the address to be used if it is necessary to report back errors. Typically, this is the originator's address, but it can be the address of a local postmaster, as it is when mail is redistributed to a mailing list.

6. **Session close**. The client requests that the SMTP server close the connection, provided the connection is not needed for other transmissions.

The following sample SMTP session shows mail sent by user *RedClaw* at *dancinglobster.com* to three people at *singingcrab.com*, one of whom is not able to be identified due to a typo in his address. Note that the following conventions are used below: client messages are denoted by a 'C' and server messages are denoted by an 'S'; square brackets enclose text that explains each event.

| Sample SMTP Session |
|---|
| S: 220 singingcrab.com Mail Service Ready<br><br>[The server responds positively to the client's session initialization request.] |
| C: EHLO dancinglobster.com<br><br>[The client initiates the SMTP connection by saying hello and identifying itself.] |
| S: 250-singingcrab.com greets dancinglobster.com<br>S: 250-8BITMIME<br>S: 250-SIZE<br>S: 250-DSN<br>S: 250 HELP<br><br>[The server responds positively and passes back the optional extensions it supports, such as responding to help queries and accepting 8-bit MIME-encoded transmissions.] |
| C: MAIL FROM:<redclaw@dancinglobster.com><br><br>[The client identifies who is sending the mail.] |
| S: 250 OK<br><br>[The server signals that the client should continue.] |
| C: RCPT TO:<harpo@singingcrab.com><br><br>[The client identifies the first recipient in the list.] |
| S: 250 OK<br><br>[The server checks that the recipient's account exists, stores the name, and indicates the client should continue.] |
| C: RCPT TO:<grouchy@singingcrab.com><br><br>[The client identifies the second recipient in the list. The user's name was misspelled and should read groucho.] |

| **Sample SMTP Session** |
|---|
| S: 550 No such user here |
| [The server checks that the recipient's account exists, finds that it does not, and replies that this user has no local mailbox. The client should return the error message to the sender.] |
| C: RCPT TO:\<zeppo@singingcrab.com\> |
| [The server checks that the recipient's account exists, stores the name, and indicates the client should continue.] |
| S: 250 OK |
| C: DATA |
| [The client indicates that it is ready to start sending the message contents.] |
| S: 354 Start mail input; end with \<CRLF\>.\<CRLF\> |
| [The server acknowledges the request from the client and prepares to receive the data.] |
| C: Date:<br>19 November 2001<br>19:05:18<br>C: From: Red Clawinsky \<redclaw@dancinglobster.com\><br>C: To: \<harpo@singingcrab.com\><br>C: To: \<grouchy@singingcrab.com\><br>C: To: \<zeppo@singingcrab.com\><br>C: Subject: Thursday Night<br>C:<br>C: Don't forget Thursday is Pasta Night!! As always:<br>C: Live Lobsters, Dancing Nightly<br>C: Maine Coon Cats, Juggling Sunday Only<br>C: See you all there.  Cheers, Red!<br>C: . |
| [The client sends the message contents and signals the end of the transmission with a dot character on a line by itself.  The headers are part of the message contents and are separated by a blank line.] |

| **Sample SMTP Session** |
|---|
| `S: 250 OK`<br><br>[The server acknowledges receipt of the message, which means it can then deliver it to the two recipients.] |
| `C: QUIT`<br><br>[The client signals the server to terminate the connection.] |
| `S: 221 singingcrab.com Service closing transmission channel`<br><br>[The server closes the communication channel.] |

When you multiply the simple transaction above by the amount of email coursing through the Internet every day, it becomes easy to appreciate the importance of doing everything possible to deliver mail as quickly and as efficiently as possible. One way this is accomplished is by requiring that the SMTP commands occur as part of a predefined sequence, like the one above. This highly prescribed process helps to reduce the time an exchange takes and to guard against errors. Also, each command has a maximum wait time associated with it, so that a connection does not remain open while waiting for a response that may never come. These and other requirements in the SMTP protocol help both the server and the client avoid wasting time and resources on bad connections or misbehaving SMTP implementations.

There are times when messages fail to get delivered even after they have been successfully transmitted. When the server SMTP acknowledges receipt of the message data, the client considers the mail delivered, unless told otherwise. But the server may encounter problems in completing delivery to the recipient's mailbox due to any number of factors. These problems may be temporary in nature or permanent. To handle such conditions, the SMTP protocol specifies that a limited number of repeated attempts at delivery should be made and, when that number is exceeded or when a permanent failure condition is encountered, the originator of the mail should be notified. This notification often includes a copy of the message that was sent and some indication of the reason for the failure. Anticipating that failures will occur and providing a course of action for the SMTP client or server to implement when they occur also helps prevent the individual computers as well as the overall network from being overwhelmed by trivial or serious error conditions, or by malicious behavior.

# Receiving and Sending Email

In the early days, for the first fifteen or twenty years of the history of email, reading and composing email was a very basic, even primitive process; and it was something you had to be taught, because there was nothing intuitive or self-explanatory about it. Email lacked all the adornments and intuitive functionality of today's graphical user interfaces. Furthermore, email did not necessarily come to you or to your computer. Before you could read your mail, you might first have to access a centralized mail server — through TELNET or some similar utility — that stored all the mailboxes for your organization. Once there, you started up the local mail program, perhaps by entering the *mail* command at a command-line prompt. What appeared on your screen was a text-only listing that displayed the sender's email address, date, and subject line for the latest ten or twenty messages. There were no buttons to click for filing, printing, or forwarding messages, nor were there any drop-down menus filled with options or colorful, helpful icons of envelopes and folders. But that didn't matter in the slightest. Once you knew the basic commands to retrieve and send email, and you could list out your messages and locate the one you were looking for, the desire to use email and the reasons for using it were the same then as they are today.

Reading email was frequently accomplished using a command-driven program that displayed the contents of your local mailbox: a plain text file containing all your emails, stored in the order in which they were received with the most recent at the top and the oldest at the bottom. Simple commands, which where often just a single character (e.g., *d* for delete, *x* for exit), were used to perform all the basic operations, including such things as listing the next or previous grouping of ten or twenty message subject lines, displaying the full contents of a specific email, and deleting an email. Sending email often entailed starting up an entirely separate mail composition program. One such program, which was standard on many Unix systems, began by prompting you with a *To:* line for you to enter your recipient's address, followed by a *Subject:* line. After you entered the subject of the message and pressed the return or enter key, you simply typed in the contents of your message, line by line. To complete the email, you entered a dot '.' on a line by itself, which signaled the mail composition program that you were finished and sent the email off to be delivered. You have seen the vestige of this dot in the sample SMTP session above. Today you no longer need to use it; but the protocol does, and for precisely the same purpose as those early mail composition programs.

You were very fortunate if the mail system you accessed incorporated some sort of program that checked spelling. Most did not, which explains one of Licklider's comments on the qualities of email as a communication medium presented earlier. But as the years went by more and more user-friendly features were added. Eventually, you were able to sort your mail into folders, automatically save a copy of your outgoing messages to a special outbox or sent folder, search messages for specific content, and even access your mail remotely using client-based programs that transparently handled the interaction with a mail server for you. Most of the features we commonly use today were originally developed and refined in those early years.

Today, email is sent and received through an email client, and the majority of these email clients use non-proprietary, openly defined protocols in order to maximize interoperability. Email clients are now included as standard software on all computers and on many Internet-ready handheld devices. Most people use an email client that is packaged as part of a larger application, such as a Web browser or a proprietary ISP interface, like America Online.

Email clients serve two fundamental purposes. First, they function as a local application through which you can read, compose, and organize your email. Their purpose in this regard is to serve as a file and information manager, providing access to your messages and to associated functions and tools, like an address book. Second, they perform the client-server operations necessary for your computer to interact with your configured mail server. Through your email client: you authenticate yourself to the mail server by sending a login and password; you connect to the server to check the status of your email (i.e., to see if you have any new messages); and you connect to the server to submit your outgoing messages to the server for transmission across the Internet. These operations are performed using one of several common Internet email protocols.

Two of the most popular email protocols are the Post Office Protocol, version 3 (POP3) and the Internet Message Access Protocol, version 4 (IMAP4), both of which are described in the following sections. These protocols, which must reside on both the client computer and the mail server, serve the same overall needs with respect to providing access to your email. But they were designed to meet different specific requirements with respect to how that access should work and where your messages should be stored.

## The POP3 Protocol

POP3 is the simpler and more pervasive email client protocol. In general, it provides for fewer operations to be performed on the server than IMAP4. POP3 confines the client-server interactions to checking for new messages on the server and then copying the new messages down to your computer. All other operations, like creating folders and copying, moving, or searching messages, are performed exclusively on your computer.

A POP3 client periodically connects to the POP3 mail server that is recorded in its configuration file. This setting for the mail server was either configured for you (e.g., by an administrator in an office environment), or it was something you configured during an installation process (e.g., following instructions from your ISP when you first opened an account). Another configuration file setting controls how frequently this connection occurs and triggers when your client queries the mail server to see if you have any new mail (e.g., every 5 minutes or every 15 minutes). During this connection, the POP3 client and server communicate in a fashion similar to the FTP and SMTP communication channels, that is through a prescribed sequence of commands and responses.

In general terms, here's what happens. The POP3 service on the server continuously listens for connecting clients on a specific TCP port. Your client establishes a TCP connection with the server on that port when you click on the button to check for new messages, or when the preset timer goes off. The server sends back a greeting to acknowledge the connection. Your client then sends commands, to which the server replies; and this conversation continues while new mail is downloaded to your computer, after which the connection is closed.

More specifically, a POP3 session proceeds in sequence through the following states:

1. **Authorization**. This state begins after the initial client-server greeting and involves your client identifying itself by sending your email account name and password to the server. Following authentication (i.e., your name and password match account information stored on the server), the server locates and locks your mailbox. Before proceeding to the next state, the server numbers and notes the size of each message. The operation of locking your mailbox is designed to prevent it from being changed by another process or session, something that might occur if two computers were trying to access your mailbox at the time. For example, you and your assistant might be checking for your new messages at the same time, or you might be logged onto two computers and both POP3 clients connect to check for new mail at the same time. Before this locking operation became commonplace (and even afterwards, when it failed for some reason), simultaneous updates to mailboxes caused grief for users and administrators alike over the years. The most common result was a corrupted mailbox file, which precipitated angry and anxious conversations over whether or not the damage could be repaired.

2. **Transaction**. This is the general command state in which information is exchanged and actions are performed with respect to listing, retrieving, deleting, and un-deleting messages. Your POP3 client issues commands in response to your actions in the email application (e.g., pressing a button to get new messages or to delete a selected message), and the mail server performs the associated operations and replies back to the client. Many POP3 mail applications include options that affect which commands can be executed in this state and what happens with the messages on the server after your client has connected. For example, you

may be able to choose to have all new messages on the server automatically transferred to your client, which means that once you are authenticated, the mail server will immediately check for new messages, copy them to your computer, and then delete them from its own filesystem. Or, you may be able to choose to have copies of your messages left on the server, forming a sort of permanent archive. Or you may be able to synchronize the deletion of your messages in some fashion so that, for example, when you delete a message locally through your client, it is then — and only then — also deleted from the server. Note that message deletion operations performed during the transaction state result in those messages being *marked* for deletion, but not yet removed. The deletion occurs in the next, and final, state. Postponing changes to the mailbox file allows for greater efficiency when making those changes and better protects the mailbox from possible corruption. Once all commands have been issued, the client issues the QUIT command so that the POP3 conversation can proceed to the next state.

3. **Update**. This is when the POP3 server actually deletes messages that have been marked for deletion, removes the lock on the mailbox, and terminates the TCP connection.

POP3 is fast and simple, which has contributed to the pervasiveness of its use. One common problem that people face with POP3, however, involves the synchronization of their mail messages. For example, if you have a computer at work and one at home and you access the same POP3 email account from each location, one of two things typically happens. Either you configure POP3 to leave a copy of all your emails on the server, or you end up with a different listing of emails on each computer. In the first instance, you have access to all of your messages from all of the computers you use; but over time your mailbox may become exceedingly large, slowing down operations on the mail server and possibly surpassing limits imposed by the mail server with respect to the size of a mailbox. In the second instance, new emails that arrive in the evening or over the weekend, and which you review from home, will not be accessible from your computer at work, nor will email you review at work be accessible from you home computer; because POP3 has copied these messages onto the respective computers. The more sophisticated IMAP4 protocol addresses this very problem, and others.

## The IMAP4 Protocol

IMAP4 is a larger and more complex protocol than POP3. It allows for much finer control over accessing and storing messages and managing email through the creation of folders on both the mail server and the client computer. The client-server communication process is generally similar to that of POP3. It begins with the server authorizing access to the client according to a supplied account name and password. It continues with the client issuing commands and the server responding. It concludes with the server performing the requested operations on the mailbox and terminating the connection. But the commands themselves are significantly different with IMAP4, because IMAP4 performs most of its operations on the mail server rather than on the client.

POP3 and IMAP4 differ most in how they interact with the individual messages you receive. IMAP4 was designed to retain your messages on the mail server and to carry out operations, such as creating folders and filing messages, on the server. By contrast, POP3's focus is on copying new messages from the server to the client and performing all other operations on the client computer. The IMAP4 email client provides you with a *view* of your messages and folders. IMAP4 also gives you the option of creating folders locally on your computer for the purpose of copying or moving messages onto your computer. But storing messages on your computer undermines the main advantage of IMAP4. By keeping your mail on the server, you are ensured a consistent and synchronized view of your messages, regardless of which computer you use to access your account. Whether you are in the office, at home, or away on vacation, you are always viewing and interacting with one collection of your messages. Also, since most mail servers are routinely backed up (i.e., files are archived nightly to protect against their loss, damage, or inadvertent deletion), using IMAP4 may lessen the risk of such loss. Most POP3 configurations result in email being stored exclusively on a personal computer, which may or may not have its local files backed up.

Whether you are using the POP3 or IMAP4 protocols, or some other protocol, the similarities in the overall process are stronger than the differences. The objective of all these protocols is to provide a common framework through which programs can connect to mail servers and then read and interpret the contents of mail folders. For most people, how easy these email clients are to configure and use and how fast they operate is all that really matters. For system administrators and technology infrastructure planners, however, the differences in these protocols have more far-reaching implications.

## Sending Email

Sending email, unlike receiving and organizing it, occurs in a highly consistent manner across most computers connected to the Internet. The communication process generally consists of an SMTP connection between your computer and the SMTP server specified by your ISP. Your email client starts up the SMTP connection and then simply copies your message to the server. The message gets queued on the server along with other outgoing messages and from there begins its journey to its destination, as described above. As far as your computer is concerned, it's that simple. It only needs to transfer your message to the awaiting SMTP server to complete its part of the operation.

# Email Message Format and Encoding

What your email message looks like as it travels the Internet is not exactly what you see in the window provided by your email client. Part of the information packaged with each message is exclusively for the mail server and client protocols that handle your mail, and this information is kept hidden from you by your email client since it has no bearing on your reading or replying to a message. Other types of information, however, are simply not shown to you, either due to the default settings of your email client or because the program was written to ignore or hide this information. If so inclined, you can view this hidden information by saving an email message to a separate file — something most email clients allow you to do — and then reading the file with a word processor. What is important to recognize is that part of every email message contains information that is generated by and included for the computers and the network, and this information is separate and distinct from that created by the sender of the message. This section describes what your email looks like from the perspective of the computers that carry it across the Internet.

Email messages are divided into two separate parts and, not all that surprisingly, these parts roughly correspond to the same two basic parts that make up the mail sent through our postal systems:

- **Envelope**. This part contains the information required to get the message transmitted and delivered. It consists of the originator's address, one or more recipient addresses, and optional information related to the protocol extensions being used by the hosts to transmit the message. The envelope is not included with the rest of the message that gets delivered to the recipient, although some of its information is also included in the message contents that does get delivered (e.g., the originator's address).

- **Contents**. This part, often called the message object, contains the information entered by the person who composed the message and additional information added by the email clients and servers that handled the message during its journey. It comprises the information that gets delivered to the recipient and itself consists of two distinct sections: a series of lines, collectively referred to as the header, that hold information about the message, and the original contents, referred to as the message body.

In the sample SMTP conversation presented above, everything shown before the client sends the DATA command relates to the email's envelope. Much like a postal envelope, the email envelope remains separate from the contents of the message and serves to communicate the information necessary to get the message to its destination. This information includes the address of the sender so that the sender can be notified if a problem occurs with the delivery of the email. The remainder of the conversation contains the email's contents: what was entered by the sender of email and what was appended by the processes and systems handling it to help track the email.

The email contents, or message object, conforms to standards in both its overall structure — its division into a header portion and a body — and in the type and syntax of the lines that constitute its header portion. This conformance is critical to maximizing email interoperability across the Internet. Each header line, for example, generally consists of a predefined field name (e.g., Subject) followed by a colon, which in turn is followed by data specific to that type of field, as shown here:

```
Subject: Aunt Rose's Secret Meatball Recipe
Message-ID: <3D8747F9.FFAA9847@qqline.net>
```

Each field name has special meaning to a mail client. Several field names, for instance, are used specifically for display purposes (e.g., the subject line), while others are operational in nature and assist the mail client in its handling of the message body (e.g., lines that indicate the language of the contents or that an attachment has been included). Some field names are required in order for an email to be minimally compliant; such compliance ensures that the message will be interpreted correctly by most, if not all, mail clients. Other field names are considered optional and may be included and interpreted by the mail clients that recognize them to refine any number and type of email features (e.g., to distinguish a new, unread email from one that has already been read). Header lines that begin with white space are read as continuation lines of the line above, which allows long lines of data to remain readable. A single, empty line is used to demarcate the header portion of the email contents from the body of the message.

Some of the information contained in a message header originates either directly or indirectly with the sender. This information typically includes lines for: the message date, who the message is from, the subject of the message, who the message is addressed to, and to whom the message is being copied (cc) and/or blind copied (bcc). Other header information is added by the services carrying the mail. As discussed above, some of this is trace information that records the path the message has taken as it passes from SMTP server to SMTP server or gateway. This information is often essential for debugging and resolving email delivery problems.

Other header lines are used by the various mail programs and servers to further identify the mail contents as it relates to their handling of the mail. These lines contain such things as unique message identification codes and reply-to lines that provide details suitable for conducting accounting operations. Many email programs allow you to view the contents of the header. By default, however, most display only a small number of the header lines, such as the date, subject, and from and to lines. But a header can easily contain a dozen or more different lines, as does the sample header shown below:

| **Sample Email Header** |
|---|
| Return-Path: <barrythecat@qqline.net> |
| Received: from mzz.lny.net (mzz.lny.net |
|    [167.206.55.22]) by mail2.zz.com (8.6/8.6) |
|    with ESMTP id g8HFPS023364 for |
|    <lobster@znax.com>; Tue, 17 Jan 2004 11:25:28 |
| Received: from qqline.net(ool-4350e345.dyn.qqline.net |
|    [67.180.227.69]) by mzz.lny.net |
|    (iPlanet Messaging Server 5.2 (built Jul 29 2002)) |
|    with ESMTP id <0H2L004J89B1D7@mzz.lny.net> for |
|    lobster@zznax.com; Tue, 17 Jan 2004 11:21:01 |
| Date: Tue, 17 Sep 2002 11:19:21 -0400 |
| From: "Barry C. Cat" <barrythecat@qqline.net> |
| Subject: Re: Lobster Madness |
| To: "J. J. Lobster" <lobster@zznax.com> |
| Message-ID: <3D8747F9.FFAA9847@qqline.net> |
| MIME-version: 1.0 X-Mailer: Mozilla 4.7 [en] (Win98) |
| Content-type: text/plain; charset=us-ascii |
| Content-transfer-encoding: 7BIT |
| X-Accept-Language: en |
| References: <3D834892.A88F7350@zznax.com> |
|           <3D835509.31329F92@qqline.net> |
|           <3D8739AF.56029393@zznax.com> |
| X-Mozilla-Status: 8001 |
| X-UIDL: 3d8749b800000001 |
| [Message Contents Starts Here] |

An email's header contains a great deal of information describing both the contents of the message and its journey. You can see at the beginning of the sample header a good example of the kind of very detailed information that gets recorded as a message traverses the Internet. The two *Received* header lines capture the names of the machines that carried the message, when the hand-offs occurred, and so on. All of this information is critical to keeping the system working smoothly and quickly. Moreover, unique identification codes embedded in these header lines function much like the bar codes used on packages. They can be used, for instance, by law enforcement or system administrators to assist in tracking down the origination point of malicious emails, or to help pinpoint technical issues that might be causing certain emails to clog the system. The following list briefly describes each of the header lines found in the sample above:

- **Return-Path**. This line holds the email address to which any return mail, such as acknowledgments or errors, should be sent.

- **Received**. These lines hold records from the mail servers that carried the message. Each line documents a hand-off of the message from one machine to the next and includes the names, IP addresses, and other associated information for the machines, and the timestamp (i.e., date, time, and timezone information) of the transfer.

- **Date**. This line holds the timestamp of when the message was sent off by the originator of the email.

- **From**. This line holds the originator's email address (which is required information) and the originator's name (which is optional information and was added by the originator's email client from one of its configuration file settings).

- **Subject**. This line holds the subject text of the message. The inclusion of "Re:", a standard abbreviation for "regarding", indicates that this message is a reply to an earlier message.

- **To**. This line holds the recipient's email address. It optionally includes the recipient's name, which the originator's email client probably added automatically from information stored in an address book record.

- **Message-ID**. This line holds a unique tracking identifier generated by the mail system that originated the message and typically includes, as is the case here, that system's name.

- **MIME-version**. This line holds the version of the MIME protocol used by the originator's email client and additional information that identifies specific information about the email client and the computer on which it was used: Mozilla denotes Netscape, 4.77 is the release number for that version of Netscape, [en] denotes that the language used was

English, and Win98 indicates that the Microsoft Windows 98 operating system is on that computer.

- **Content-type**. This line holds information about the message contents and here indicates that the message is plain text and encoded in the US-ASCII character set.

- **Content-transfer-encoding**. This line holds MIME-related information about any encoding that was performed on the message contents before it was sent so that the corresponding decoding can be performed by the recipient's email client. 7BIT indicates that the default US-ASCII encoding was used, therefore no decoding is necessary.

- **X-Accept-Language**. This line holds one or more language identifiers that specify the language of choice for any replies to the email's originator. English is identified here with en.

- **References**. This line holds unique identifiers that associate one message with one or more other messages. It is typically used to preserve the connection between an original message and one or more responses that include or reference that original message, as occurs when an email client's Reply button is used.

- **X-Mozilla-Status**. This line holds message status information for a specific email client (e.g., Mozilla lines are inserted and used by Netscape email clients). Different codes are used, for example, to identify whether a message is new, if it has been read, or if it has been marked for deletion. If the code indicates that the message is new, for example, this might cause the email client to display the message in boldface text to distinguish it as unread. 8001 indicates that the message has been read.

- **X-UIDL**. This line holds a unique identifier used by the POP3 protocol for retrieving mail from a server.

One piece of information you should not see in a header is the name of a recipient who has been blind copied on the message. The blind copy header line (i.e., bcc) requires a small bit of special processing in order to keep those recipients hidden. Three different implementations to handle this special processing are described in the email message format specification. The first calls for the bcc line to be removed before the message content is distributed to the list of recipients. The second calls for the bcc line to remain in the message sent to those blind copied but removed from the message sent to the other recipients. The third leaves the bcc line for all recipients but removes the address information for any individuals being blind copied; this lets everyone know that some people were blind copied without actually identifying them.

Following the header, the body of an email message simply consists of lines of text. These lines, like those in the header, are supposed be kept to 78 characters or less in length in order to optimize viewing for traditional displays that were standardized on 80 characters per line. Lines can, however, be up to 998 characters in length. In RFC terminology, the 78 character limit represents a SHOULD requirement while the 998 represents a MUST requirement.

Equally important, from the perspective of Internet interoperability, is the requirement that the contents of email messages must only contain characters from the US-ASCII character set. This basic character set handles plain text English messages without any significant restrictions, but it is wholly inadequate for other languages (e.g., Japanese or Hebrew) and for transporting any type of non-text data, such as photos, drawings, and audio files. This character set restriction posed problems long before today's world of multimedia files. But it's fair to say that the ubiquitous nature of email owes a lot to this restriction (or simplification). A larger character set, or a looser specification allowing for different character sets, would have greatly inhibited the growth of email and probably resulted in more messages being lost or corrupted. Consider your own email experiences. Think about how reliably email is delivered and try to recall the last time you received a message that was damaged in transit.

The solution to overcoming the restriction imposed by transporting ASCII-only content centers on using a program to re-encode all non-ASCII content into the conforming ASCII character set prior to its hand-off to SMTP. Email attachments are one result of this process. No matter the type of file, its contents will be re-encoded into ASCII and marked as an attachment (by means of additional header lines) in preparation for the file being sent as part of an email. Many mail programs can also encode and decode non-ASCII messages on the fly, which is particularly useful if you are communicating in something other than English. The specifics of this process are discussed in the following section.

# MIME: Turning Email into a Full Service Information Carrier

Multipurpose Internet Mail Extensions, or MIME, transformed email from a simple mechanism for transporting basic text messages into a highly standardized content delivery system that could handle any and all types of files containing any type of information content. Regardless of the type of data or the character set used, MIME specifies exactly how to identify and package data so that it can be carried across the Internet by means of the existing mail protocols and not become corrupted in the process.

More specifically, MIME was created to redefine the format of mail messages in order to enable the following capabilities:

- Using character sets other than ASCII in the body of mail messages. This allowed email to be composed in any language, not just those few languages that could be adequately represented by the limited character set of ASCII.

- Using character sets other than ASCII in the message header. This further facilitated the use of other languages, removing yet another restriction that forced the Internet to operate in its restrictive, English-only mode.

- Allowing message bodies to contain a large variety of non-textual formats. This meant email could potentially carry any type of file, and existing standards could be used for how to represent many of these different file formats.

- Allowing message bodies to be divided into identifiable parts. This enabled email to comprise different elements. For instance, one part of an email might contain standard message text while another might contain an attachment with a photograph or application file.

The single most critical aspect of MIME is that it extended the email specification without introducing any incompatibilities. It accomplished this in part by introducing new header fields. These new header fields functioned to further classify the type of message being transmitted (e.g., the message content was Japanese and a JPEG file was included as an attachment). This information, in turn, signaled the need for specialized processing by mail programs and mail handlers (e.g., encoding the message contents into ASCII before its transmission across the Internet to protect against corruption of the data, decoding it once the message was received, and launching a program capable of displaying the JPEG file). These header fields include the following:

- **MIME-Version**. This field indicates that the message is MIME-conforming and allows for different versions of the MIME specification to be accommodated. (The absence of this header line is used to identify older, non-conforming messages.)

- **Content-Type**. This field specifies the media type and subtype of the message body, thereby characterizing the data so that a mail reader can render it appropriately. For instance, media type might be set to *image* to indicate a graphics file, while subtype might be defined as *jpeg* or *gif* to indicate the specific file format. Any number of file types, existing now or yet to be invented, can be accommodated in this header line.

- **Content-Transfer-Encoding**. This field shows how the message data was encoded before being transmitted, so that the associated decoding mechanism can be employed when the message is read.

- **Content-ID and Content-Description**. These fields further describe the message data contents.

Note that some of these header fields are included in the sample email header presented in the previous section, even though the message itself is plain ASCII text. The additional header lines serve to better describe and identify the mail contents. The more exacting the information contained in the mail's header, the better the mail clients can determine how to present the message to you. Sometimes this involves doing little or no additional processing, as is common with simple ASCII text messages. But, other times, the mail client must decode message contents before displaying them, or it must interact with your computer's operating system to check for the existence of a particular application or a certain type of character set. It's this behavior and these tasks that MIME helps to define and enable.

Five principal data content types have been defined by MIME: text, image, audio, video, and application. Any existing type of data can be accommodated by indicating which general category the message data falls into and then specifying its subtype. In this way, MIME even allows text-based messages to be further refined into something called enriched content. Enriched content is what allows email messages to include typical text processing features, such as font designations and character sizes. Moreover, plain text type messages can also indicate their particular character set, allowing the demands of non-English languages to be met.

MIME defines two additional composite media types, called multipart and message, in order to handle email that includes a mixture of data types and encapsulated messages, respectively. Multipart messages are actually quite common, although most mail clients now handle the distinctions between data types so seamlessly that multipart emails don't appear to consist of several distinct parts by the time they are displayed to you. But any time two messages of different data types are combined, as happens for instance when you reply to a message and quote the original, this process creates a message with a multipart content type.

The message content type, on the other hand, is frequently used for very large emails that, because of their size, need to be broken into pieces and transmitted separately. Because this type is an encapsulated message, it can contain partial messages that can later be recombined. The concept is similar to how IP handles individual datagrams or packets which are then later combined and reassembled at the destination through TCP. Often this message type is used for transferring large application or operating system files via email. Since many systems have restrictions on the maximum size of email messages, breaking up these large files into segments overcomes this restriction and also provides a more secure way to transport such a large amount of data. For instance, if a problem is encountered and one or two messages fail to arrive, only a fraction of the whole file needs to be retransmitted, rather than the entire file.

Since many of these data types cannot be represented by the ASCII character set in their native format, a content-transfer-encoding must be specified in the header. This header line shows both the encoding mechanism — such as base64 or quoted-printable — used to translate the native file format into ASCII and an indicator (or domain) of its native format (e.g., binary or 8-bit). Both pieces of information assist the mail client that accepts the message in determining which process needs to be followed to undo the encoding (i.e., to decode) before you can view or work with the message contents.

MIME is responsible for converting the early Internet email of plain ASCII text into today's multifaceted, multimedia email that carries text in any language and can transport such things as photos, audio and video files, and application files from word processors and engineering programs.

# Email Uses, Abuses, and Concerns

As explained above, the basic mechanics of email are relatively simple. Email clients effectively hide the complexities and file details from us, while MIME safely packages and identifies the contents of email, and the protocols of SMTP, POP3, and IMAP4 quietly and efficiently handle the transportation of email to and from our computers and across the Internet. But many issues relating to our use of email and to how email is used in the business world are more subtle in nature. Moreover, the distribution of unsolicited email and the sometimes malicious intent of its content demand that we become and stay more informed about how email is being abused in order to remain safe and secure in our own use of email. The following sections explore several of the more prominent issues related to email, such as the use of mail forwarding and mail aliases, the hazards of unsolicited email, and concerns over email privacy and security. Several of the issues presented below are explored at greater length in "The Technology Revolution."

## Mail Forwarding

Mail forwarding is the practice of configuring one email address as a public front to another email address, and is similar in effect to the service of mail forwarding offered by the postal system. The address to which email is being forwarded, moreover, is often kept deliberately hidden, sometimes due to concerns over privacy and other times due to business implications or simple logistics, as is explained below. This practice is becoming more and more common on the Internet as mail forwards (i.e., the name for the public email address) are being used by individuals and organizations for an ever widening variety or purposes. The following list describes a few of the most common uses:

- You are migrating from one email address to another and, like moving from one home to another, forwarding enables mail to automatically travel on to the new address rather than be returned to the sender.

- You want to establish a single, permanent address that will stay constant regardless of which ISP you use or which company you work for. A mail forward will allow the address people use to reach you to remain the same, whether you change jobs or ISPs frequently, infrequently, or never.

- You want to keep your true address less visible on the Internet in the hope of avoiding solicitations or other forms of junk mail. A mail forward may enable you to remain less conspicuous and may even assist in the process of filtering unsolicited email so that you never have to see such messages.

- You want to use a general, or generic, address that represents some facet of your business (e.g., sales@twincats.com), but you don't want the additional expense or overhead of having mail stored at that address and accessed there. A mail forward will redirect any incoming mail to the general address to the email address of a specific individual assigned to handle such correspondence.

A mail forward operates as a simply relay: mail arriving for the public forwarded address is accepted locally and then sent on to the associated destination address. The sender of the email may or may not be made aware that the message has been forwarded to a different address. That's because SMTP can be configured to treat mail forwarding in one of two ways. In one configuration, the forwarding is kept silent as the SMTP server reports back to the sender that the mail has been successfully delivered, even though it must still forward the message on to its final destination and it may fail in the process. In the other configuration, the forwarding includes a return notification process that sends a message to the originator of the email with information about the updated address. This way the sender can update their information. Both approaches have their uses and their drawbacks, and selecting one over the other depends on the reasons for using mail forwarding in the first place.

But the biggest problem with mail forwarding, which is growing in relevance as its use continues to climb, is that it defeats a fundamental and essential feature of email, that of delivery verification. In the event of a failure with the forwarding of a message, the sender will believe that their message was delivered and the intended recipient will never know that anything was sent. Meanwhile, more and more mail will accumulate as "dead letters" on intermediate mail servers.

## Mail Aliases and Lists

Mail aliases and lists are two types of pseudo-mailboxes, that is, special types of email addresses that are defined by SMTP and, unlike other addresses, are not associated with any local user account. Mail aliases and lists provide a different type of mail forwarding service, one designed specifically to simplify the management of an organization's email by enabling email to be sent to one address and automatically redirected to one, or more than one, specific individuals. More than anything else, it is their uses that differentiate mail aliases from mail lists. Many corporations and other types of large organizations use mail aliases to manage external emails arriving from customers and mail lists to better facilitate internal email communications.

A common use of mail aliases is to establish generic mail accounts to handle unsolicited email from customers (e.g., for sales or technical support). Such aliases serve to hide the identities of the individuals performing these functions and, in doing so, they simplify the task of assigning and reassigning responsibility for reviewing and responding to any messages sent to those addresses. Such reassignments are commonly needed, due to illness, vacations, or rotations in staffing; and mail aliases localize the process of reassignment to a quick and simple change to an email configuration file. Most importantly, how the customer communicates with the organization remains unaffected by these configuration changes. For instance, an alias for the address *support@dancinglobsters.com* allows incoming mail to that address to be directed automatically to one or more other users (e.g., the local users *bigpete* and *littlepete*). The SMTP server accepts the incoming mail and verifies its delivery, as the protocol requires, keeping *bigpete* and *littlepete* hidden from the sender behind the alias. It then sends each of the petes a copy of the message.

Mail lists also work by taking a single pseudo-address and redirecting incoming mail to one or more email addresses that make up the associated list. A mail list, however, is more a redistribution of email than the simpler type of email forwarding done through an alias. It is primarily used to expand one generic address (e.g., salesteam@dancinglobsters.com) into a list of individual recipients (e.g., the addresses for all fourteen members of the sales team). The mail list reduces the effort required to send an email to all the team members by enabling one address to be specified rather than all fourteen. Like an alias, the mail list hides the individual email addresses that constitute the list from the sender, allowing any changes to the list to remain transparent to the sender. One technical difference between mail lists and aliases is that the return address used for errors with mail lists — unlike the case with aliases — is changed from the sender's email address to that of the local postmaster who administers the lists. This is done because SMTP assumes that such errors are generated by problems with the list itself; returning such information to the sender will not correct the problem, because it is the list itself that needs fixing.

Mail aliases and lists are simple but powerful features of SMTP. Both provide a highly useful layer of abstraction, separating one or more local email accounts from general-purpose email communication functions, such as providing generic email aliases for customers to use or establishing a company-wide email list name for broadcasting emails to every employee in the company, all management-level employees, or all members of the technical support group.

## Unsolicited Mail: Chain Letters, Viruses, and Spam

Email, like other Internet activities, produces its share of unwanted, undesirable, and even unethical behavior. As soon as you establish a presence on the Internet with an email account, you become a potential target for abusive behavior. People of all ages can be victims of various types of unsolicited email activity. But children should be a special concern and require the greatest amount of protection. The most common forms of unsolicited email, chain letters, viruses, and spam, are briefly described below. "The Technology Revolution" explores these areas in more detail, with particular attention to the security and privacy of our information and, more generally, our presence on the Internet.

Chain letters distributed through email on the Internet have shown themselves to be as, or more, damaging than their postal equivalents. Email chain letters are hoaxes that rely on a relatively small number of recipients unwittingly recirculating these hoaxes to their friends, family, and business associates who in turn do the same thing. The result is the generation of a lot of unnecessary email traffic and wasted space in countless mailboxes, in addition to whatever scam may be described in the chain letter itself. One of the most difficult problems in combating chain letters is that they can spread faster and farther on the Internet than information that exposes them for what they are and tries to minimize their impact and curtail their distribution.

Hoaxes propagated through chain letters appear in a variety of different forms. One particularly malicious type contains a seemingly helpful warning about a quickly spreading virus and includes details on actions you should perform to protect your computer from becoming infected. Following the instructions in this type of hoax typically results in some type of damage to your computer's operating system or one or more of its applications. You, in effect, become the instrument that does the harm. Other types of hoaxes include those that tempt people with stories of free merchandise, scare people with warnings about products or activities that may or may not actually be harmful, prey on people's sympathies by trying to get them to send money to someone in need, and spread urban myths with horror stories describing gruesome, but unlikely events (e.g., a cat or dog exploding in a microwave). These Internet hoaxes are often compelling reading. What's worse, they may often be forwarded from someone you know, which gives greater credibility to their content. You may even feel compelled to pass along one of these hoaxes yourself, unwittingly contributing to the problem. Fortunately, there are sites on the Internet that document these hoaxes and try to limit their impact, such as the U.S. Department of Energy's Computer Incident Advisory Capability site located at http://www.ciac.org. So, when in doubt, always find another source to confirm or deny the message; and if you don't know the sender of such an email, be immediately suspicious.

Computer viruses that are spread through email carry a greater personal hazard to your files and computer than most chain letter hoaxes. The effects of a computer virus can range from something innocuous, designed to demonstrate the originator's cleverness, to something very harmful, designed to destroy files on your computer or to damage its operating system. Most viruses travel concealed inside, or as, an email attachment (i.e., a separate file included by the sender of the email that is distinct from the message contents of the mail). An attachment must be opened by a program on your computer in order for its contents to be viewed; and usually, but not always, this requires a separate action on your part (e.g., clicking on the attachment) before it will be opened. Microsoft Word files, for example, are frequently sent across the Internet as attachments in email and have, for that reason and others, become the file type of choice for many creators of computer viruses. Attachments that contain executable files (e.g., files with a .exe filename extension), however, are potentially the most damaging type of computer virus files and should be avoided at all cost; they can do virtually unlimited harm to your computer.

You might think that knowing the sender of an email would be sufficient reason to lower your guard and open an attachment. Unfortunately, many viruses are passed from friend to friend and family member to family member without anyone realizing it until it's too late. If in doubt, simply don't open the attachment and delete the message. The virus can do no harm if the attachment is never opened. Locally installed virus protection software will also help locate and remove such threats before they harm your system or you inadvertently pass them along. They are not, however, nor can they ever be, a complete and foolproof solution, given that there will always exist some lapse between the time new viruses start spreading and when this type of protective software can be updated with new information.

Unsolicited junk email (or spam) is another problem altogether. Sadly, the more time you spend on the Internet, the more mail you generate, the more sites you visit, the more forms you fill out in which you provide your email address, the more likely spam will find its way to you. Your best overall protection is to limit where you provide your email address and to be especially careful about replying to unsolicited emails that offer to take you off of their list. This can provide confirmation of your identity and, unfortunately, may have the opposite effect of what's desired and stated. Spam filtering software also offers a limited means to automatically categorize email as spam, (e.g., by comparing the sender's address against a known list of spammers and matching certain key words in an email's subject line) and delete it before it can become part of your listing of new mail. But a comprehensive solution to spam will require new, enforceable legislation, because only legislation can stop the distribution of spam at its source.

It's unfortunate that the same qualities that make email so popular — its ease of use and the speed with which it delivers information — are precisely what make it so suitable for being exploited by means of chain letters, viruses, and spam. In general, the best defense you can make against being victimized by unsolicited email is to become better informed about the damage it can impart and remain current in your understanding of its hazards and in any protection you may have chosen to combat it (e.g., email filtering and virus scanning software). Always err on the side of caution when reviewing any new email, especially when it's from someone you don't know and, even more so, when that email contains an attachment. Thankfully, there are far more people on the Internet working to protect your security and privacy than there are people abusing the Internet's openness. If you haven't already, learn to use the power of the Internet to your advantage. Visit sites that track emerging threats and vulnerabilities on the Internet, such as the US Computer Emergency Readiness Team (US-CERT) at http://www.us-cert.gov.

## Privacy and Security

Few people think about the privacy or security of their email while composing messages to friends or even while exchanging business email that contains confidential information. But as more and more people use email, as more and more business transactions are conducted through email, and as more and more organizations rely on email for much if not most of their internal and external communications, email privacy and security is fast transforming itself from a peripheral concern into a core requirement for next-generation email systems.

Two obstacles stand out above the rest when examining the current email protocols from the perspective of the privacy and security of the information they carry. First, SMTP itself is inherently insecure. Authentication mechanisms and other basic security features are sorely lacking in SMTP. Armed with even a superficial understanding of the protocol, an individual who is so inclined would not find it difficult to negotiate a connection with an SMTP server directly and create messages that seem to come from someone else. This type of email fraud is referred to as *spoofing*: sending email that claims to have originated from one source, perhaps a trusted friend or business institution, but was instead sent from another source altogether. Second, email is sent as plain ASCII text; this means it is readable. Anyone who has access to the filesystem that stores your email, or to the SMTP server filesystems transporting your email, can read it easily and without your knowing about it.

Any email system that promises to guarantee the privacy and security of the messages it carries will need to integrate the following two features:

- **Authentication**. This feature will ensure that the sender of the message has been verified through some secure and tamper-proof methodology or process to be the person and/or organization represented by the return address of the email.

- **Data Encryption**. This feature will secure the contents of a message through an encryption process that will make the contents unreadable by anyone but the message's intended recipient.

The technology already exists to authenticate the sender of an email (e.g., through a verification mechanism called a digital signature), and data encryption software has existed for decades, ever since the first modern computers were built in the 1940s. Moreover, private email systems used by some businesses and governments today routinely deliver and verify private and secure email messages. The problem is that these additional email features, as available today, are just that, additional, and require greater effort than most people are willing or have time to expend.

It is only a matter of time before safe, secure, and trusted email will be commonplace on the Internet. For this to happen, these new features will need to be implemented in such a way that we hardly know they are there. But product integration and, more importantly, general interoperability (the hallmark of the Internet) have so far eluded those trying to bring private and secure email to everyday users of the Internet.

## Web Mail

The World Wide Web introduced a different way to read and send email on the Internet, one that removed the need for an independent email client to be installed on your computer and functions without either the POP3 and IMAP4 protocols described above. A Web browser is all that is needed to access and use a Web-based email account, and, accordingly, the communication that occurs between your computer and the Web server while you read, compose, and manage your email is controlled by the Web's principal protocol, HTTP.

Some people are confused by this distinction between Web-based email and the email conducted through email clients. That's because their Web browser and email client are typically packaged together as part of a single "browser" application, as they are in Netscape's Communicator product and Microsoft's Internet Explorer, and this has fostered the impression that all of their interactions with the application are in some way dependent on or associated with the Web. But the Web browser and email client in these applications are separate and distinct components. Moreover, the email client has no dependence on or connection to the Web whatsoever. The application windows for each may look similar in terms of their colors, icons, and the common components that can be accessed from drop-down menus, like an address book or spelling checker. But their similarities are superficial, not functional. The email component relies on SMTP and uses either POP3 or IMAP4; and it interacts exclusively with an SMTP mail server to send off your messages and to retrieve new mail. The browser component, on the other hand, interacts exclusively with Web sites and relies on HTTP. It makes no distinctions between a Web page displaying the latest news stories and one listing your email messages.

Accessing email through the Web and HTTP has certain advantages over accessing email through an email client and POP3 or IMAP4. One benefit of Web-based email is that there are no geographic limitations on your sending and receiving messages. Web-based email can be accessed from any location that provides basic Internet access. By contrast, if you connect to the Internet from more than one location or through more than one ISP, you may not be able to access the Internet email account provided by your ISP or business from each and every location. Another benefit with Web-based email is that, other than needing access to a Web browser, no configuration settings are required on the local computer, unlike email clients which need to record settings such as which protocol you are using (e.g., POP3 or IMAP4) and the name of the SMTP server used by your ISP. This means that you can easily access your Web-based email from a friend's computer or from a public library, something that would be difficult or impossible to do with email accessed through a client application.

The disadvantages of Web-based email are related more to performance than functionality. The basic operations of viewing and composing messages are generally slower than through an email client. This is because your email is stored exclusively on the Web server, as opposed to being copied to your computer. Also, because most Web-based email accounts are free, advertisements are printed alongside your email in order to offset costs for the service, and this adds to the time needed to load each page. Features vary from site to site with respect to the types of operations you can perform through any particular Web-based email account. Most, however, do not offer as many features for managing and organizing your messages as come standard with email client applications. Another potentially large disadvantage of Web-based email has to do with the privacy of your email and, more generally, your use of the site. Most Web sites are in the business of selling information about their users, and this may impact, for instance, the amount and type of unsolicited email you receive.

A common use of Web-based email is as a secondary account for people who use computers at their workplace. The data that resides on workplace computers is supposed to be entirely work-related and, as such, is considered the property of the company or institution that owns it. In other words, there is no such thing as personal email at work. The owners of the workplace have the legal right to read every piece of mail you send or receive; and, technically speaking, this is extremely easy for them to do. Using a Web-based email account, however, allows you to read and compose personal emails and avoid introducing any personal data onto the local computer or the local network; the email resides exclusively on the Web server and no mail files pass through the office mail systems for anyone to copy, read, or complain about.

Web-based email provides a different way to access and manage email. It does not, however, alter or otherwise impact any of the other components of email on the Internet. For example, the transmission process (i.e., getting email from its point of origin to its destination) remains unchanged and is still handled by SMTP. The composition of email messages also remains unchanged: header lines store information about the message's sender, its intended recipient(s), its contents, and its journey; and MIME helps to categorize the type and packaging of the message's data.

## An Email Aside

The following email is a good example of one of the many email inventions or stories that circulate from time to time for purposes of amusement or shock. It illustrates the point that it's not always the technology that fails or misleads; often it's the people.

A couple from Minneapolis decided to go to Florida for a long weekend to thaw out during one particularly icy winter. Because both had jobs, they had difficulty coordinating their travel schedules. It was decided that the husband would fly to Florida on a Thursday, and his wife would follow him the next day.

Upon arriving as planned, the husband checked into the hotel. He decided to open his laptop and send his wife an email back in Minneapolis. However, he accidentally left off one letter in her address; and, without realizing his error, he ended up sending the email to an entirely different address.

In Houston, a widow had just returned from her husband's funeral. He was a minister of many years who had been 'called home to glory' following a heart attack. The widow checked her email, expecting messages from relatives and friends.

Upon reading the first message, she fainted and fell to the floor. The widow's son rushed into the room, found his mother on the floor, and saw the computer screen which read:

To: My Loving Wife
From: Your Departed Husband
Subject: I've Arrived!

I've just arrived and have been checked in. I see that everything has been prepared for your arrival tomorrow. Looking forward to seeing you then! Hope your journey is as uneventful as mine.

P.S. Sure is hot down here!

# Cyberspace: The Internet as Meeting Place

## Cyberspace: A Brave New (Virtual) World

William Gibson coined the term cyberspace in his futuristic, 1984 science fiction novel, "Neuromancer." His definition of cyberspace reads:

> A consensual hallucination experienced daily by billions of legitimate operators, in every nation, by children learning mathematical concepts...a graphical representation of data abstracted from the banks of every computer in the human system. Unthinkable complexity. Lines of light ranged in the nonspace of the mind. Clusters and constellations of data. Like city lights receding...[1]

Gibson's vision of a complex network of computers interconnecting billions of people across the globe predates by more than a decade the kind of pervasive computer use and widespread Internet access that originated in the latter half of the 1990s and is common today. The fictional particulars that compose the cyberspace of his futuristic world are not what come to mind when most people think about cyberspace. Yet the cyberspace given form and function by the Internet and the cyberspace envisioned by Gibson in Neuromancer are one and the same with respect to their basic composition: a global network of computers that functions to interconnect people, machines, and information and that has acquired the form of a virtual space through which people can navigate and interact with others.

Unlike other man-made (and woman-made) creations, there is no physical substance to the objects that make up cyberspace or to its encompassing environment. Instead, cyberspace is composed entirely of information. It is an utterly artificial construct, one made possible by the interconnection of computers and derived exclusively from the information they contain. Anything that can be described and categorized can take shape and become an object in cyberspace; this facet of cyberspace is precisely what makes it so diverse, compelling, and empowering (and so difficult to define). Accordingly, cyberspace can include recreations of real-world objects, like an exacting three-dimensional representation of a twenty-story glass and steel office tower that is part of the New York City skyline; and it can just as easily include products of the imagination, like a menacing representation of a thirty-foot-long fire-breathing dragon taken from the pages of a novel written by J. R. R. Tolkien.

It is these information objects that constitute cyberspace. They make up the contents of the environment by defining and giving form to such things as furniture, rooms, buildings, streets, animals, and people; and they establish and regulate the conditions that control how the environment operates by defining such things as how to hold a private conversation as opposed to one that can be overheard or interrupted, which rooms are locked and which are open, how fast one can travel from place to place, and whether or not one can create new objects or alter existing ones. Together, these objects, their associated conditions and features, and their encompassing environment compose a form of virtual reality in which and through which thousands of people routinely interact across the Internet.

Consider a room as an object to be described for its inclusion in some cyberspace environment. The room can be defined for the computer in terms of its dimensions, whether or not it has windows (and if it does, where those windows are located in the room, how big they are, and how they open), where its door is located, the color of its walls and ceiling, and the material used to cover its floor. The room — itself an information object — can also be described in terms of the objects it contains: a desk made of oak with four drawers and a filing cabinet, a crystal vase placed on the desk that contains a single flower, eight inches in length with six blue petals, an arm-chair situated in front of the desk, five more arm-chairs placed around a glass-topped, round conference table in the corner of the room, and a whiteboard hung on the room's far wall.

Once the room and its contents are defined as objects by entering information about their form and function, a computer can create any number of different representations of the room's space. For example, it might render the space in the form of a three-dimensional image, complete with its contents. It might additionally allow the room to be viewed from different perspectives, perhaps by simulating someone walking from the room's doorway to the whiteboard at the far side. It might also allow the objects in the room to be moved about or altered (e.g., applying a different paint color to the walls or changing the type of wood used in the furniture). Similarly, descriptions of ourselves and depictions of fictional characters can take the form of objects that reside in and contribute to the composition of cyberspace. This is, for instance, what enables individuals to enter the room, sit in the chairs, and write on the whiteboard. None of these objects have physical mass, but in the digital world of computers that fact is irrelevant. All the information a computer contains is captured and stored as some sequence of zeros and ones, so why not a representation of an office building, a room, a desk, a dragon, or even us.

Even though objects in cyberspace have no mass, each object does possess physical traits. There are measurements that describe an object's physical characteristics, such as its length, width, and height. There are also features that describe an object's physical qualities, such as the scent given off by a flower, the information or paraphernalia stored in the drawers of a desk, or the speed with which a dragon can fly. There are also traits that describe an object's behavioral characteristics, such as whether or not a dragon is afraid of cats or if a squirrel prefers one type of nut over another. Once objects are defined, a computer can collect this information together to create a spatial metaphor that gives form and substance to cyberspace. The metaphor created might be exceedingly simple, like the rectangular terminal window that is commonly used to represent a chat room; or it might be complex, like the richly detailed cityscapes built for virtual reality gaming environments. The purpose of these spatial metaphors is to create a container for the information objects that also functions as an interface through which we can interact with those objects. The room and its contents described above would have no value if they were not part of a context or setting where individuals could meet and interact. Think of cyberspace as the encompassing environment for all of these various spatial metaphors — a meeting place in which all these digital, information-based objects can reside, co-exist, and interact.

This chapter explores the ways in which the Internet has become just such a meeting place — an environment in which people not only exchange information, but also gather and interact. Since the environment of cyberspace consists of information objects, and there are no limits on the type or quantity of objects that can be created or how these objects can be made to interact, people may choose to play or work in cyberspace in any number of ways. The following sections describe three of the most popular ways in which people routinely interact in cyberspace: chat, games, and virtual reality simulations.

## Chat: Talking in Cyberspace

Chat represents the simplest type of cyberspace interaction over the Internet. In its purest form — there are many different kinds of chat programs and chat environments — chat allows two or more individuals to exchange text-based messages in real time (i.e., live or synchronously). This exchange of information takes place in a chat room, which is a cyberspace construct for a shared space. The chat room, which appears as a window on each participating individual's terminal, embodies a virtual meeting place where individuals can gather and effectively *talk* through the exchange of messages entered through their keyboards.

Internet chat, like email, holds its own distinctive place in terms of helping people reach out, connect, and communicate over the Internet. On the surface, the two services seem to have a lot in common. Both email and chat are highly popular forms of communication and are used on a daily basis by millions of people. Both allow people to take part in private, one-to-one communications, or one-to-many group communications. Both are used for personal communication and have also been successfully adopted for use in business environments. Both are accessed and used through free and easy-to-operate tools. Both enable people to utterly ignore geographic constraints in order to share information with family, friends, business associates, and others. They accomplish all this, however, in strikingly different, although complimentary, ways.

Upon closer examination, Email and chat actually have little in common; this is true of their basic technology (i.e., how they function) and it is true of how and why they are used. Email over the Internet is fast, but asynchronous. It lacks any true sense of immediacy — any interactive give and take — even in the fastest of email exchanges. Email messages are written and replied to with little or no regard for when those messages will be read, just like letters sent through the postal system. Additionally, email correspondence is often collected and saved. It is filed into folders for safe keeping, sentimental reasons, business purposes, or simple reminders, also just like letters.

In contrast, Internet chatting is immediate and synchronous. Chat resembles talking over the telephone more than it does exchanging email or letters. Chat conversations are spontaneous in character; as such, they are generally informal in tone and frequently abrupt in manner. The immediacy and fast-paced nature of the interactions allows little time to collect and compose one's thoughts and demands a more truncated style of communication, one filled with its own peculiar and rich vocabulary of abbreviations, acronyms, and jargon (see appendix A for examples). The immediacy of chat also impacts the subject matter of its discussions as much as the style and quality of its discourse. Whether chatting with a co-worker across a room, a friend a thousand miles away, or a group of strangers distributed across the country, the focus of a chat discussion is entirely on the here and now. Once the conversation ends, which very often happens abruptly, it's like putting the telephone down after saying goodbye. The conversation is closed, for the moment anyway, and normally no record is kept regarding the information that was exchanged.

Programs for sending synchronous text messages through computer terminals date back to the early 1970s and were used over the years for a variety of purposes, ranging from informal one-on-one conversations to complex business conferences with dozens of people in multiple locations. The simple, but central concept of these early text-messaging computer programs was to make possible real-time communication among groups of two or more people. The Internet was the ideal medium for this type of communication to flourish, as it has with chat.

The networked, Internet chat commonly used today is essentially a text-based group communication system that exhibits all the best and worst qualities of CB radio, a cocktail party, a conferencing system, and a party-line phone system (i.e., a group of people sharing one phone line). In very general terms, an individual starts up the chat program installed on his or her computer (chat programs are often included as part of a browser application), examines a listing of chat rooms that are open and the types of topics being discussed (chat rooms are often categorized by their subject matter or their intended audience), and selects a room to enter. A plain, rectangular window opens on the user's terminal to represent the shared space of the chat room; it is typically divided into two areas. One area displays the contents of the conversation from all the members in the room; the messages scroll by one line, or a block of lines, at a time, prefaced by the identification of the person making the comment. The other area accepts text for the user to enter and then send into the room.

Another common form of Internet chat is one that consists of a private conversation between two or more individuals. Chat programs typically include a feature (e.g., buddy lists) that enables users to build a type of address book to hold the identities of the people they frequently communicate with through chat. Through this feature, or by conducting a search, a chat program can discover whether or not a particular person is currently online and, if so, it can create a chat room for the exclusive use of those two individuals. A window opens on each person's terminal to represent the chat room and enable the exchange of messages exclusively between those two individuals. They also have the option of inviting others to join them in their private chat room.

Many other, more specific types of chat environments also exist. The technology puts few limits on how to configure a chat-based environment in order for people to interact and exchange information. Chat has been used, for example, to extend online technical help to customers in place of or in addition to telephone-accessible assistance, to conduct virtual business meetings, and to enable college professors to hold open office hours for students without requiring either the professor or students to visit the office.

Several different types of chat architectures have been developed since the late 1980s. The three most common types are:

- **Internet Relay Chat (IRC).** This is the oldest and most widely used chat architecture, with more than 100,000 people online and chatting at any given moment. IRC is an open-source, feature-rich, client-server version of chat that is used throughout the world to bring people together to discuss common interests, or simply to provide an informal means for individuals to gather privately and chat. The following section explains how IRC works.

- **Instant Messenger Chat.** This is a proprietary form of chat architecture that is essentially a simplified and specialized version of IRC developed by various corporations for their specific audiences. The most popular implementations are AOL Instant Messenger, Yahoo!Messenger, MSN Messenger, and ICQ (a play on "I seek you"). These implementations offer free client programs that are easy to configure and use. In order to use them, however, all users must have the same client program on their computers.

- **Webpage Chat.** This type of chat architecture occurs through a Web browser and a hosting Web site. It does not require that a chat client program be installed on the user's computer. Typically, a Java applet (a type of mini-application) is loaded by the user's browser and operates transparently as the client, allowing the user to see the chat rooms that are available and participate in the conversations. Webpage chat is frequently built upon an IRC chat environment, conveniently allowing IRC to handle the chat conversation traffic while using the Web to display the results and accept input from the participants.

Common to all the different types and uses of Internet chat is how chat creates a spatial metaphor — most typically referred to as a room — that allows people to meet and talk in cyberspace, no matter where they may be physically located. The simplicity of chat, both in how the client programs operate and how the various systems were engineered and function, belies the rich and profound nature of its impact. Chat may represent the simplest form of cyberspace interaction, but the dynamic and immediate nature of that interaction is tangible and compelling. Chat offers more than a simple, rectangular window through which to exchange messages. It turns the remoteness and immense expanse of the Internet into something familiar and friendly. Chat makes it possible to build a neighborhood of one's own from a collection of chat buddies, making it easy to locate and meet up with others across the vastness of the Internet and, just as important, making it easy to be found and contacted.

## The Mechanics of Internet Relay Chat

IRC enables people all over the world to gather in public or private virtual rooms called channels and converse together over the Internet in real time by typing messages on their computer keyboards. The original IRC program was written in 1988 by Jarkko Oikarinen at the University of Oulu in Finland. He wanted to improve on the existing utilities that allowed users of online bulletin boards to talk to one another while connected to a particular bulletin board service. His program spread quickly across the Finnish and Scandinavian networks. It then hopped over the Atlantic and spread rapidly across North America.

Today, IRC is documented in four separate RFCs (RFC 2810 — RFC 2813). These RFCs define the overall architecture of IRC, the management of chat rooms, and the protocols that reside on the client and the server that handle the computer-to-computer negotiations and the transfer of chat traffic. In order to use IRC, an IRC client program suitable for the particular computer's architecture and operating system must first be installed. The client is normally configured to automatically locate the nearest available IRC server, but it also provides the option to connect to other servers that may offer a different or wider selection of chat rooms and topics. Once the client and server negotiate a connection, the user is able to choose a unique nickname to use and select a chat room to enter.

It was the Gulf War of 1991 that brought IRC to the public's attention. For a full week after radio and television broadcasts from the region were cut off, live updates describing the events that were transpiring came across IRC connections. Many IRC users gathered on a single channel (one virtual room) to hear the reports. The information conveyed in these chat rooms was cited in newspaper, radio, and television reporting on the Gulf War. Similarly, during the attempted coup against Boris Yeltsin in September 1993, IRC users in Moscow provided live reports of the events as they were occurring.

IRC operates on a client-server model. In general, the servers collect and distribute the messages throughout the network, while the clients establish connections to local servers and enable users to send and display the messages. For each channel (or room) the servers echo incoming messages to all the users logged into that channel. This copy and echo process occurs so quickly across the Internet that it provides the illusion that everyone is interacting in the same virtual room.

An IRC network typically consists of a group of servers that talk directly to one another. Clients connect to one server and from there are able to join any channel on the network that is known to the local server. While it may seem like clients are talking to each other (particularly when only two people are chatting), clients only communicate with the servers, never directly with each other. When a message is sent, it goes to the local server the client is connected to, the one configured in the client's IRC program. This server then passes it along, as needed, to the other servers that make up the network. When a message from another chat room participant is displayed, however, it may come from any other server on the network, not just the locally configured server.

Unlike other data transport systems, which route data efficiently from node to node, each IRC server must be able to send messages to every other IRC server in order to deliver messages as quickly as possible. Additionally, each IRC server must be aware of all the network-wide channels that are in use and who is logged in to each channel. As this information changes — as people come and go and channels open and close — the servers must synchronize this information in order to maintain consistency across the network. Channels can be established that are local to one server or to a subset of the network's servers. Such channels will only be seen and, consequently, can only be joined by clients connecting to those servers.

Each channel contains at least one person with special, authoritative control. This person is referred to as an operator or *ops*. Depending on the channel, this person may simply be the first person to open the channel, or it may be someone who was assigned this responsibility, as is frequently the case in more formal channels. In any event, the operator is supposed to monitor behavior and has the ability to kick users off the channel and even to prevent them from returning.

Channels can also be private. Two or more people can establish an unadvertised channel and talk privately together without being concerned about being interrupted or overheard. They can also invite others to join the channel by sending out a notification to specific users. Many people save money on their phone bills by communicating in this manner.

## The Business Side of Chat

IRC is an inexpensive, effective alternative to face-to-face business meetings. Some people refer to meetings over IRC as non-meeting meetings. People can gather for a conference or a brain-storming session while remaining at their desks. Inexpensive IRC software packages designed for business use make it easy to create private, password-protected channels and assign one individual as the moderator (or ops) to set a meeting's agenda and help keep discussions focused on the topic at hand. Chat-based meetings eliminate many of the time and geographic constraints that can keep a conventional meeting from occurring. Additionally, some people who might be less inclined to speak up and make a contribution in a conference room filled with fifty or more people may find it easier to express themselves by typing in their comments in a chat environment. IRC can also record the entire session, providing complete minutes of the meeting. This means that no one needs to be designated as the meeting's secretary or scribe; and the minutes for the meeting can be distributed to all the participants, including those who might have joined late or had to leave early, and to others who could not attend, immediately after the meeting's close.

In addition to business meetings, IRC is used to manage interactive customer support environments. These environments often provide real-time customer assistance in place of or in addition to a help-line telephone number. IRC is also being used to conduct interactive training sessions of new employees. A trainee's skills can be developed by allowing them to interact with other individuals posing as customers in a chat environment; a trainee's product knowledge and social skills can also be tested and evaluated in this manner. Educational institutions are also using IRC in what has come to be called *distance learning*. IRC provides the means to establish classrooms in cyberspace, in which teachers and students can ask questions, offer answers, and otherwise interact and exchange information; attendance can be taken, too.

## The Personal Side of Chat

For most people, chatting over IRC is as simple as starting up a chat client, selecting a room to enter or contacting a buddy, and entering into conversation. Interacting in the cyberspace of Internet chat rooms augments rather than replaces other types of communication and face-to-face meetings. For others, IRC is a world unto itself and is compelling to the point of being truly addictive. One indication that IRC use can become dangerously habit forming for some people is the existence of a USENET newsgroup called alt.irc.recovery. The newsgroup serves those who are unable to refrain from chatting and who are in search of help and support to assist them in reducing the time they spend in chat environments.

The habit forming aspect of chat is not its only potential danger. As is true of many other services on the Internet, IRC is largely uncensored; as such, it exhibits the same good, bad, and ugly characteristics that people and society display outside of the Internet (or, what is called in cyberspace, IRL — In Real Life). The atmosphere in most chat rooms, whether filled with familiar individuals or with strangers, can be friendly, entertaining, and educational, or abusive, mean-spirited, and offensive. Fortunately, it's safe to say that the majority of chat rooms are friendly, inviting, and polite environments. But there is no one checking identity cards at a chat room's door. No person or policing agency is in place to confirm an individual's age, or sex for that matter. It is not uncommon for people to pretend to be someone they are not. Children in particular need to be protected from the potential dangers that such misrepresentation can create (although the same can be said for unsuspecting and trusting adults, too).

## The Rules of Chatting

Chatting, like other activities on the Internet, has become and remains popular due largely to the fact that the majority of people obey some basic, clearly stated rules with respect to their behavior in cyberspace. But people who do not obey the rules, whether stated (as below) or implied, should understand that chatting is not the anonymous activity many people believe it to be. The fanciful nicknames selected for chat mask one's true identity as far as others in the chat room are concerned, but chat log files automatically record information identifying the computer and network connection being used by each participant. This information is sufficient to discover the true identities of most individuals.

The following sample list of chat rules is fairly representative of most chat environments:[2]

1. Offensive behavior is not allowed in any channel. This includes harassing other users by pestering them, calling them names, racist remarks or swearing. Our channel operators will kick or ban offending users at their sole discretion for offensive behavior.

2. Obscene nicknames are not allowed in any channel. Our channel operators will ask offending users to change their nickname. If the user does not comply, they may be kicked or banned at the sole discretion of the channel operator.

3. Flooding in a channel or in private will result in a user s connection being terminated from the server. If a user persists in flooding, they may be banned by a channel operator or k-lined from the server.

4. Users are not permitted to connect to this chat server through a proxy server or firewall. The use of such connections will result in a z-line or k-line.

5. Users are not permitted to place bots in any channels. The use of bots will result in a channel ban.

6. We do not tolerate hacking, nuking, cracking, or any type of DoS attacks on or by our users. We take this very seriously and will assist federal and local law enforcement in prosecuting offenders. Threatening users in the manners mentioned will result in an immediate k-line.

Chat rules like those specified above serve to define the overall policies regarding expected behavior and punishable misbehavior in any one chat environment. As these rules typify, chatting has developed a vocabulary all its own. But the meaning behind all the jargon is not difficult to discern (e.g., z-line and k-line are technical terms for banning someone from the service); and it does not take long to fully understand what others are saying or to start using the jargon oneself.

# Games: Playing in Cyberspace

Email transformed the Internet from a technology that connected computers into a network that connected people. Internet chat took that network of people and added to it an immediacy that was lacking in email by enabling two or more people to gather in a virtual room and hold a conversation. The playing of games across the Internet took the immediacy of chat and the nascent social composition of cyberspace one giant step further. Games introduced the element of virtual or simulated environments onto the Internet along with a framework for interacting and sharing experiences in those computer-generated environments. In doing so, games added a whole new dimension to the Internet's network of computers, transforming the network into a playground and forever changing the way in which people can meet, play, and socialize.

At one time, the act of gaming referred exclusively to the practice of gambling. But computers introduced another definition for gaming, one that focuses on the playing of games that simulate conditions and allow one to role play. Because computers can be programmed to simulate any type of object, setting, environmental factor, or action that can be described and categorized, they can create gaming environments with virtual representations that approximate the physical world, animate wholly imagined worlds, or simulate anything in between. Accordingly, popular gaming environments have been created to simulate such things as war zones, boxing matches, and medieval worlds of dungeons, ogres, and magic, and to allow role playing that includes such actions as firing a rifle at enemy troops, throwing punches at an opponent, and casting evil spells.

This type of gaming, however, with its emphasis on virtual reality recreations, reaches far beyond the world of entertainment. Simulated environments and the simulated behavior they make possible work just as well for the purposes of work-related testing and training exercises, such as the training of pilots with flight simulators or the testing of police cadets with crime-scene simulators. (Many computer simulation programs that were created for work-related training have been repackaged for the gaming marketplace, often with few if any modifications.) Computer simulation programs have also been adapted to suit the purposes of self-paced education. These programs sometimes appear in the form of a game (e.g., a program that teaches children math skills through puzzles); they also take the form of computer-driven tutorials.

What all of these programs share, whether designed for entertainment, training, or education, and whether played with others across a network or by oneself on a single computer, is that they use a computer system's resources and its store of information objects to simulate a dynamic environment that responds to and changes in some measure based on a player's movements and interactions within that environment. The integration of networking into these programs, however, and especially that of the Internet, resulted in something more than a simple enhancement to how these programs could be accessed or how many people could use them at the same time. Networking was a catalyst that transformed the industry and culture of gaming, and it made possible entirely new applications for the simulated environments that these games and virtual reality recreations contained. This, in turn, changed the composition of cyberspace. Cyberspace became capable of shaping the broad, formless landscape of the network into a specific and recognizable destination, such as a richly imagined world of dungeons and dragons or, more simply, a chess or monopoly board. These destinations quickly became popular places where many people choose to meet, talk, interact, and play.

The following sections present how gaming has impacted the evolution of cyberspace. The first section describes the unique and lasting contributions made by two of the first computer games. It's followed by an explanation of networked role-playing games generally referred to as MUDs, which have become the most common form of cyberspace gaming environments on the Internet, and a short section on how traditional games have been adapted for play across the Internet. The final section focuses on the mechanics of interactive gaming; it describes the different types of network architectures used in providing gaming services and the various network factors that influence the performance of gaming.

## Early Computer Games

Individuals interested in computers very often demonstrate an affinity for gaming. One reason for this is that computers and gaming are a natural match. At a fundamental level, games consist of predefined, static rules coupled with variables that are introduced by the players and change as the players progress through the game. The same holds true for computer programs. A program's code consists of rules that define the commands, options, and working environment (e.g., the look and feel of the program's window or interface) available to the user, which in turn determine what a user can and cannot do; and the user provides all the dynamic, or variable, content (e.g., the text for an email or the shapes and colors for a drawing).

Two of the earliest computer-based games were Spacewar! and Adventure, which date back to the early 1960s and early 1970s, respectively. They served to spark interest in the use of computers for gaming, which at the time was something far removed from how and why computers were generally used and, more basically, how computers were regarded. These games became popular, however, because they were fun and exciting to play.

Widely considered the first interactive computer game, Spacewar! was created in 1962 on one of the earliest minicomputers, a DEC PDP-1. Because of its compact size, the PDP-1 could be programmed and used directly by an individual, unlike most other contemporary computers; such direct access to a computer was a prerequisite for playing any type of interactive computer game. By contrast, most computers of the time were very large, complex devices that few people were allowed to interact with directly. These computers — commonly called mainframes — resided in separate climate-controlled rooms that were specially designed to house all of their components. Access to these rooms and to the computers themselves was typically restricted to a small number of trained technicians. Moreover, programs written for the computers were executed by proxy and asynchronously through the batch processing model (i.e., handing off a program to be run by a computer center technician and waiting for results from the program to be returned in some printed form). The PDP-1 minicomputer, therefore, represented a radical departure in terms of basic computer use, programming, and accessibility. It seems only appropriate that one of the first sophisticated programs written for the PDP-1 — an interactive gaming simulation — would itself represent a radical departure in terms of the type of operations a computer could be instructed to carry out.

Spacewar! was created by a group of computer enthusiasts in Boston (most of whom were students at MIT) with the PDP-1's scant nine kilobytes of memory (typical inexpensive personal computers sold today are equipped with several thousand times this amount of memory) and a thirty-line display. The game allowed each of two players to control a spaceship, one of which was called Wedge and the other Needle. The spaceships were programmed with limited fuel supplies and the ability to fire torpedoes in an effort to destroy the opposing spaceship. A sun positioned at the center of the display simulated a gravity well that exerted a pull on the spaceships proportional to their distance from it, making the game that much more challenging. The backdrop of stars was a faithful representation of one section of the night sky, complete with many familiar constellations and with the relative brightness of the stars programmed from astronomical data.

In terms of its engineering and programming, Spacewar! was less a game than a serious, complicated, and demanding science project. Building it relied on a working knowledge of mathematics, physics, and astronomy, in addition to a detailed understanding of computer programming, memory allocation (there was precious little of it), and even the characteristics of the computer's display. Spacewar! was a feat of computer engineering in the guise of a game (something true of many of the computer games that would follow). Copies of Spacewar! circulated quickly at universities in the U.S. during the 1960s and 1970s and, understandably, it found its way onto some of the earliest hosts of the ARPANET. Versions of Spacewar! can still be accessed and played on the Internet today, and it stands out as an early indication of how important gaming would become to the continuing development of computers.

For many of the people who saw or played Spacewar!, it was far more than an amusing and challenging game that sparked interest in the use of computers for recreation and gaming. The PDP-1 and Spacewar! changed the way they fundamentally viewed computers by presenting a powerful and entirely new paradigm for computer use. This effect was not incidental. The original purpose behind the creation of Spacewar! was to build an engaging demonstration program that would both show off the PDP-1's capabilities and tax its resources. Spacewar! far exceeded these expectations. Most of all, it impressively demonstrated for the first time an element of computer-user interaction that today is taken for granted: that of an individual interacting directly with a computer. This interaction just happened to take the form of a simulated environment containing information objects displayed as spaceships, stars, and torpedoes.

The game of Adventure (sometimes referred to as ADVENT, because many computers in the 1970s and 1980s limited filenames to six characters) was created ten years after Spacewar!. It presented a very different type of gaming experience and, accordingly, made a different contribution to the development of computer gaming and, more importantly, the future composition of cyberspace. Adventure was the first computer game to present an interactive, virtual world. Unlike Spacewar!, which was a game of action with competition between two players in a fast moving graphical environment of spaceships and stars, Adventure was a puzzle. It pitted the player directly against the computer and the intelligence of the program in an effort to map and navigate safely through its mysterious world of interconnected caves. Accordingly, Adventure was a game of wit, reason, language skills, and imagination.

The creation of Adventure has direct ties to the building of the ARPANET. Its creator, William Crowther, was working at BBN in Boston in 1972, developing the assembly language program used by the IMPs, the early ARPANET routers. While there, Crowther combined his interests in caving, rock climbing, and a role-playing game called Dungeons and Dragons in a computer game he wrote for his daughters. The game incorporated a rudimentary simulation of his cave explorations along with fantasy elements of role-playing. Since he was creating the game for his children rather than other computer programmers, he designed the game to be directed by the player through natural language statements rather than through rigid, static commands that a player would first need to learn. This feature alone set the game apart as something new and exciting.

Adventure was essentially a game of exploration. The program defined the objects that constituted and inhabited Adventure's simulated world; these objects included a stream, various rooms that were part of interconnected chambers in a cave, treasures, a bird, and locks and keys. The program also provided the framework for playing the game, which controlled how the player interacted with Adventure's simulated world, learned about it, and moved through it. The game was played simply by entering text in a window, such as "walk upstream" or "open door;" and the game replied with one or more simple, declarative sentences, responding to the player's input or describing the existing game environment. For example, anyone who has played Adventure will recognize the following description:

You are in a twisty maze of passageways, all alike.

As the game was played, more and more of the hidden, simulated world of Adventure was revealed. But progress was gradual, and extensive notes and maps needed to be created as the game was played if one were to avoid the dangers and discover the secrets that Crowther had cleverly hidden. That was the challenge and, in a word, the adventure.

Crowther wrote Adventure in FORTRAN — the first high-level computer language — for a DEC PDP-ll computer. The simulated environment was based on caving maps made by him and his wife of the Mammoth and Flint Ridge cave systems in Kentucky. Many of the actual cavern names (e.g., "The Hall of the Mountain King" and "Twopit Room") were also included. Specific features of the caves are part of the program, as is extensive use of caving terminology. The program traveled far and fast on the ARPANET; it was copied from computer to computer, often in the middle of the night. Not surprisingly, it quickly became responsible for many a sleepless night, as network administrators and researchers gained access to the game and began their journey through the maze of passageways and awaiting surprises that filled Adventure and made it so entertaining and challenging.

Adventure was later enlarged in 1976 when Don Woods, working at Stanford University's Artificial Intelligence Lab, came upon the program and took an interest in developing it further. He located Crowther over the ARPANET and secured his permission to make changes. Woods, like many computer people of the time, was enamored of the fantasy world portrayed by J. R. R. Tolkien in "The Hobbit" and "The Lord of the Rings" trilogy. So, among other elements, Woods introduced a troll, elves, and a volcano into the world of Adventure. Soon afterward, the game was rewritten in the C programming language, became widely distributed across the Internet on Unix machines, and was even included on the first IBM personal computers.

Even though Adventure pitted the player against the game's programming, rather than against another player, and there was no network involved in playing the game, it illustrated to many people the as yet untapped potential within the computing environment for creating and exploring simulated, or virtual, realities. The player moved through the gaming world of Adventure, manipulated objects, triggered events, and interacted with inhabitants, thereby changing the game's virtual reality by becoming part of it. The use of natural language statements, like "free bird," "cross river," and "unlock gate," heightened the immediacy of the player's interaction with the simulated environment. The player effectively came to inhabit the game's virtual world, learning more about that world and becoming more familiar with it as he or she progressed through the game.

Adventure marked the beginning of something far bigger and more engaging that its twisting passages, dangerous creatures, and golden treasure suggested. It was for many an early introduction to the realm of cyberspace. The only thing lacking was the ability to share this experience with others. When networking was later integrated into this virtual environment for game playing, allowing multiple players to interact within a gaming world from anywhere across the Internet, a new, more dynamic, social gaming experience was created, as the next section will explain.

## Networked Role-Playing Games: Multi-User Dungeons (MUDs)

Before networking was introduced into the world of gaming, computer games had already acquired a large and passionate following. The personal computer revolution of the 1980s brought easily as much gaming software as productivity-based software (e.g., word processors and accounting applications) to the shelves of computer stores. But computer games, like other types of games, are best when played with others. They are an inherently shared experience, with personal interactions and competition contributing to the fun, excitement, and difficulty of play. Networking provided the means to enlarge the shared experience of gaming.

The growing access to computer networks in the 1980s, therefore, was viewed by gaming enthusiasts from a slightly different perspective than that shared by most other network users. While the others focused their attention on remote computer access, email, and the transfer of data across the network, gaming enthusiasts focused on how a gaming environment and the actions of individual players might be shared across the network. This thinking led to the development of entirely new types of gaming environments, the creation of which added a new dimension to the Internet's burgeoning realm of cyberspace. These new games, unlike the games that preceded them, recognized the transforming qualities of the network itself; they incorporated and relied on the network as a fundamental component in their gaming environments.

The first real-time Internet gaming experience for multiple users was named MUD, which stood for Multi-User Dungeon. MUD was written in 1979 and 1980 by Roy Trubshaw and Richard Bartle at Essex University in England on a DEC-10 mainframe. MUD provided a game not unlike Adventure, but one that allowed for any number of simultaneous players connected over a local network and over the Internet. The game consisted of a series of interconnected chat rooms in which players could not only talk, but they could also move about and interact with surrounding objects in the game world and with one another. An experimental packet-switched network of computers connected Essex University to the ARPANET, allowing people from Europe and the U.S. to access the game. This, combined with computer magazine articles about MUD in the early 1980s, generated a great deal of interest in MUD, which in turn inspired many people to write their own multi-user games. The unanticipated and widespread popularity of MUD caused Bartle and Trubshaw to form MUSE Ltd., where they rewrote their game into MUD2. (The term MUD has since come to refer to this genre of multi-user role-playing games and has acquired the more generic meaning of Multi-User Dimensions.)

MUDs have become a thriving form of entertainment and social interaction on the Internet. Thousands of different MUDs have been created. They vary considerably in terms of their content, style, and conditions of play, and cater to a wide variety of interests and audiences. Most MUDs, like the original, are fantasy, role-playing adventure games in which players traverse the game world's environment alone or in groups and participate in solving puzzles, sleighing monsters, or journeying on some quest. These MUDs are often grouped and categorized according to common themes. Themes may be general in nature, such as historical MUDs that simulate the world of ancient Egypt or the 19th century American frontier. Or themes may be highly specific in their content, such as MUDs that recreate a world from fiction, like that of "The Lord of the Rings" or of "Harry Potter", or that simulate a world from a popular television show or movie, such as "Star Trek" or "Star Wars." The majority of these MUDs are new and imaginative fantasy world creations, designed and built by an individual or a small group, and complete with their own history, laws, gods, monsters, currency, buildings, and landscapes.

Other MUDs are oriented less towards action-based exchanges in cyberspace and more towards intellectual pursuits and person-to-person communication. MUDs of this type resemble an interactive newsgroup and focus on the exchange of ideas and the pursuit of more traditional socializing. The constructed cyberspace environment, its objects, and the actions one can perform within it, are of secondary importance. Such MUDs are often differentiated by a specific interest (e.g., literature, movies, or current events), a particular location, or an age group. These MUDs are virtual communities that more closely resemble chat rooms; people go primarily to talk with other players, but they can also move about, use virtual objects, and create new objects and rooms, in order to augment their interactions, entertain one another, or just about anything else. Someone might, for example, create a virtual deck of cards so that a group could play poker or bridge as they talked.

MUDs vary considerably in terms of how they are played. Many MUDs are free and can be accessed through freely available client software, while others are fee-based and may require proprietary software to be loaded onto the player's computer in addition to a fee to start playing the MUD and/or an hourly, monthly, or yearly recurring fee to keep the account active. MUDs can be strictly text-based, like the earliest role-playing games, with all interactions occurring in a chat-like window. Players enter commands to move about and to interact with the environment, and send private or public messages to the other players; the actions and messages from other players, or those generated by the game itself, are also displayed in the window. MUDs can also be graphical, with images of the MUD environment and other players rendered by the MUD's client program. Graphical MUDs enable players to view the game world through their *avatar*, the character that has been created for their role playing, enhancing the effect of interacting in cyberspace.

In terms of their operation, interactive multi-user games such as MUDs rely on a client-server architecture not all that different than the architecture used for chatting via IRC. The client-side component consists of a program that must be installed on a player's computer; it comes configured to exchange messages with one or more MUD servers. Alternatively, some text-based MUDs can also be accessed through TELNET, as was done with the first MUDs from the 1980s. A player begins by selecting a unique identifier, just as with chat, and then defines his or her role or persona (i.e., their avatar) by selecting from various characteristics stored in the server's database. These characteristics vary considerably from MUD to MUD and may include such common factors as age, sex, height, weight, and type of hair, and less common factors such as whether one is human or some type of forest animal and whether or not one possesses special powers or skills. For instance, a separate classification exists for anthropomorphic MUDs; all players must choose some non-human animal form and then apply characteristics and traits — animal and human — to the animal form they have chosen to represent themselves in the game. Each player interacts with a MUD through his or her client program by engaging other players they come into contact with, interacting with the environment, or interacting with characters and responding to events generated by the MUD itself. In addition to sending public or private messages, players can also express themselves non-verbally by *emoting*. Just as chat conversations often include graphical or text-based emoticons to communicate such things as a smile, kiss, or smirk, MUD clients include a command for doing such things as storming out of a room, laughing, dancing, or hugging another player.

MUD servers define the characteristics of the virtual world, typically by means of objects (e.g., rooms, furniture, and weapons) and properties (e.g., what the rooms, furniture, and weapons look like and if they have any special powers) stored in a database. The database also stores information about the players, their chosen characteristics, their activities, and what they can and cannot do. Additionally, the server contains programming logic to respond to each player's actions during the game; the logic imposes rules on what a player can and cannot do and keeps the game moving. Like a chat server, it must also relay player messages, actions, and any other gaming events to all the players so that everyone's game world remains synchronized.

Because a MUD is essentially a framework of discrete objects and the MUD server is responsible for keeping track of all of these objects, each player's movement or action in the framework entails a series of related operations and responses on the MUD's server. For instance, a simple action like one player passing through a doorway and entering a room will require the MUD server to perform a series of operations in order to update and maintain the environment for all the players. These operations may include:

1. Identifying the object representing the player who issued the command, the objects representing the room he or she was in before, and the objects representing the room he or she has entered.

2. Deleting the player from the property list of the old room.

3. Adding the player to the property list of the new room.

4. Checking for objects in the old room that represent players and sending each of them a message that this player has departed.

5. Checking for objects in the new room that represent players and sending each of them a message that this player has entered.

6. Sending the player the property description of the new room.

MUDs are highly dynamic environments. Like the real world, they are changed by events that transpire within them (e.g., a flood, famine, or fire) and by the actions of their inhabitants (e.g., one player killing another, the construction of a new building, the discovery of a treasure map). This feature alone is responsible for making MUDs lively and popular cyberspace destinations that many people visit frequently. Also, just as in the real world, players change over time and may be directly or indirectly affected by other players, specific events, or changes in the environment. It is common for different players to be given different capabilities. Some players, for example, may be given the authority and the tools to build onto the game world by defining new rooms, objects, and features, thereby enlarging the virtual world. Other players, by comparison, especially novices (commonly referred to as newbies), may only be able to interact with the environment, move from room to room, handle objects, and exchange messages.

MUDs may be generally categorized as games, but they are not toys. Most MUDs, for example, contain thousands of unique rooms (a room is a general MUD term for any single object in the environment in which players can move and interact, and may describe such things as a house, a store, a cave, or a forest). In addition to the basic contents of a MUD — the objects that define and fill the virtual environment — each MUD defines its own rules and customs (and sometimes entire histories) that serve to refine the unique quality and character of the MUD and guide and enforce behavior within it. The creative and technical effort required to build a MUD is considerable. The vast majority of people building MUDs have been, and continue to be, programming enthusiasts who have demonstrated considerable skill in pushing the limits of what computer programming is capable of achieving. Through their imaginations and a compelling desire to create virtual environments that are at once new and challenging, many MUD creators have built interactive environments that are as much serious artistic creations as they are feats of engineering.

Interacting in MUDs across the Internet is a very popular form of recreation, and not just among teenage boys eager to battle for domination of this, or some other, planet. MUDs offer a wonderfully diverse and creative outlet, a place where friends and strangers alike can role play, build, alter, and destroy worlds, and interact within adventures of their own making. There are no limits to the types of virtual environments that can be created, whether for role playing, more traditional socializing, city building, historical recreations, or even for business applications, as is discussed later in this chapter. Nor are there limits on the types of persona one can create, in order to remain anonymous and perhaps act more adventurously than one feels capable of, or comfortable with, doing in the real world. Moreover, unlike chat rooms and newsgroups, the objects one creates in a MUD — a weapon, a building, an entire city — remain there for others to discover, use, or even alter in one's absence. All of these factors combine to instill a sense of permanence and dimension to these virtual worlds that seems to reach beyond the programming and databases that make them possible.

In one form or another, all of these MUDs are constructs for social interaction in the Internet-wide, computer-based, information-bound realm of cyberspace. The network overcomes the constraints imposed by geography with respect to meeting and interacting with others. Computers provide the means to create the virtual environments in which to meet and interact and the interface through which to access the virtual space and extend oneself into it. Information stored on the computers and transmitted across the network defines the objects, events, behavior, and communication that together constitute each MUD's virtual world, its inhabitants, and the players' interactions with one another and with the world itself. MUDs are responsible for giving form and substance to cyberspace that is unrivaled by any other type of Internet service or offering. What makes them engaging, powerful, and popular Internet destinations is actually something simple and basic. It has to do with the desire for people to meet and interact with others, which at times is coupled with the need to play.

## Traditional Gaming Adapted to the Internet

The commercialization of the Internet in the early 1990s accelerated the effect of networking on computer gaming, much as it affected other common uses for computers, such as email, remote computer access, and information sharing. The effect of the Internet on computer gaming was not, however, limited to the creation of new types of gaming experiences, like MUDs. Traditional games, along with gambling, also found a place on the Internet; and they made their own distinctive contributions to the composition of cyberspace.

The playing of board games (e.g., monopoly, scrabble, checkers, and chess) and card games (e.g., pinochle, hearts, and go fish) are traditional and familiar social activities. Each of these games is defined by its own unique set of rules, and each requires specific objects (e.g., a board and pieces, or a deck of cards) in order for it to be played. When personal computers became popular, software versions of these games were created that allowed individuals to test their game playing skills against the computer or against other players gathered around the computer. The Internet brought these games into the realm of cyberspace.

Network versions of these games enable friends and family to gather by means of the Internet in order to interact, have fun, and compete with one another when meeting in person might not be possible. The Internet provides the common ground — the meeting place — and the games define the particulars of the environment, such as the layout of the board, the pieces used by the players, the wheel to spin or the dice to throw, and the rules by which to play. Playing these games in the cyberspace of the Internet does not need to replace face-to-face gatherings, nor can it ever hope to come close to the intimacy of playing a game with everyone seated around a kitchen or dining room table. But these Internet game versions offer a familiar and entertaining way to stay close and connected when circumstances (and life in general) get in the way.

A different type of traditional gaming that has made its way onto the Internet involves gambling, or gambling-type games. These types of games have also acquired a considerable following across a diverse group of Internet users. They constitute the Internet version of the most common types of betting games that have been played for many years. They include the following basic game types:

- **Casino**: blackjack, roulette, slot machines, poker, etc.

- **Lotteries**: daily, weekly, instant cash, scratch-off, and pull-tab games.

- **Sportsbook**: wagering on the outcome of professional sporting matches.

- **Pari-Mutuel**: wagering money into a pool which is then divided amount the winners; bettors wage against each other, not the house.

This type of gaming is chiefly done through the World Wide Web. The games rely on the Web's protocols, its ease of use, and its graphical page layouts to simplify matters of presentation and to handle the game's interaction with each player. In general, the network is used simply to provide remote access to a graphical representation of the same gaming mechanisms that can be found in Las Vegas, a newspaper stand, the football pool at work, and so on.

A number of Web sites are dedicated to gambling; that is how they make their money. Others use these games, either played for fun or for money, as incentives to get people to visit their site and, more importantly, to get these same people to return. Even when not played for money, the majority of these games (with the possible exception of poker) are less about social interactions than the excitement of gambling. Cyberspace is not always a place one meets other people. Sometimes it is simply a destination, where one can, for instance, spin a virtual roulette wheel or buy a lottery ticket.

## The Mechanics of Interactive Gaming

The mechanics of gaming in a networked environment are not all that different from the basic mechanics of other network services, such as chat or the Web. On one level, these mechanics are all about the transmission of data and the intercommunication between the local computer and some remote computer on the network. But the large amount of data being transmitted and the speed requirements of the interactions in gaming equates to greater demands being made on computer system resources and on the network.

Real-time interactive gaming across the Internet stretches the limits of the type and quality of services that can be offered over the Internet. Given the open nature of the Internet, there are no rules or requirements that individuals or companies must adhere to in order to offer gaming on their sites. This, combined with the fact that many games are graphical in nature and fast-paced, means that interactive Internet gaming typically produces a lot of network traffic and can therefore consume a considerable, and often disproportionate, amount of the general network resources that are available (compared to other popular activities, like email and Web browsing).

Everyone shares the general network resources provided by the Internet. But there is no overriding notion of equality or fairness that ensures an even or equitable distribution of these resources; nor was any engineering done or protocol features defined (for instance, in TCP/IP) to implement some control over the distribution of network resources based on the type of usage or a particular computer or user; nor is the cost for using the Internet tied to the amount or type of data an individual sends or receives. Information regarding how much or little an individual consumes of the large, but finite network resources that are available remains hidden from the individual user. This information, however, is not hidden from network service providers (e.g., ISPs) or from others monitoring network usage. In order to deliver the best overall service for the most users, some ISPs strictly prohibit certain types of interactive Internet games, either forbidding their use altogether or limiting access times. The same holds true for many businesses that provide their employees with Internet access. Firewalls and other network devices allow companies to control the types of services that are available; and this type of selective access to Internet services is likely to increase.

While the Internet's resources cannot be allocated based on how they are used or by whom, there are network factors that can be controlled by the individual and the network providers that will impact the performance of playing games and other network-intensive activities. There are also different types of communication architectures used in providing gaming services over the Internet, each of which exhibits different pros and cons with respect to gaming performance and the use of network resources. The following sections briefly examine the types of network factors and communication architectures that affect the performance of playing games over the Internet.

## Network Factors that Affect Gaming

Many different factors affect the overall performance of any computer game, starting with the hardware of the computer itself (e.g., the amount of memory and the type of central processing unit). Gaming over the Internet introduces several additional factors that may affect a game's performance. These factors and their impact vary considerably based on the particular service or provider, the type of network connection (e.g., dialup or broadband), and the type of game (e.g., a text-based MUD or a graphical war simulation). The following three factors exert the greatest influence over gaming performance:

- Network bandwidth

- Network latency

- Computational power

These factors correlate to physical properties of the network that define the limits of what gaming can and cannot do across the Internet. Bandwidth relates to the capacity of a network connection, or pipe, for transmitting data. The larger the pipe, the more data can be simultaneously transmitted. The smallest pipe is likely to be the one between a person's home and their ISP. For most people this connection is over a phone line via a modem; but more people are opting for a broadband connection via cable or satellite to obtain, as its name suggests, broader bandwidth. Interactive games with graphical interfaces demand some type of broadband connection in order for them to function reasonably well across the Internet. The largest pipes connect major hubs along the backbone of the Internet. For instance, a T3 is commonly used to connect major hubs and has a bandwidth of roughly 45 million bits per second, while a 56K modem on a standard home computer handles at best 56,000 bits per second, and typically far less.

Bandwidth is affected by the size and frequency of the game's messages (i.e., the commands and communication to and from each player) being transmitted across the network as well as by the number and distribution of users. Also important, especially in the context of gaming, is the technique used to distribute messages. With MUDs, most messages are meant to be delivered to multiple recipients, with each recipient representing a different player. How the server is configured to handle this one-to-many distribution impacts network performance. As these games scale up, with more and more players joining the network, the traffic generated increases exponentially. As more bandwidth is consumed by gaming, less is available for other services, like email, FTP, or viewing Web pages.

Network latency is the delay factor incurred as data is transmitted from one node to the next across any network. In general terms, latency has continually decreased over the years as more bandwidth has been added, network protocols have been refined, router technology has been improved, and the wired infrastructure in many places has migrated from slow copper phones lines to fiber optic cable with its nearly speed-of-light transmission rate. There will, however, always be some delay factor. The transmission of electrical signals take some finite amount of time no matter how small or large the distance; and delays will be added by packet routing, queuing, and processing.

Different amounts of network latency are acceptable for Internet gaming; it all depends on the type of game being played. Highly interactive, fast-paced games such as those that have combat simulations demand very low latency amounts. Otherwise, the gaming environment will have a jittery, rather than fluid, appearance. Other game types, like many MUDs and board games, can operate successfully with higher network latency.

The computational power of the servers used to control multi-user interactive games will also affect perceived performance by the players. Just as there are limited resources in the network, each server has limited disk space, memory, and processing power. How each server is configured and monitored is critical to a game's performance. This is particularly true as the number of players increases, which causes server resource demands to rise exponentially. Part of this configuration relates to the network architecture employed for the gaming environment. These architectures are described next.

## The Network Architectures of Internet Gaming

Network and server resource limitations will always exist. This is as true for gaming purposes as it is for other Internet services. But gaming can make significantly higher demands on these resources. One way these resources are maximized is through the selection and configuration of a distributed network architecture that best meets the particular gaming demands. The following communication architectures are employed for multi-user Internet gaming:

- Peer-to-peer
- Client-server
- Server-network

A peer-to-peer architecture consists of a direct connection between players; this means that there is no intermediate host performing any type of filtering or control. Each player's computer directly broadcasts or multicasts (i.e., transmits) its messages to all the other connected players. This type of approach works quickly and efficiently when there are a small number of players, or when players are interconnected on a local area network rather than across the Internet. There is, however, no way for this architecture to grow and adapt as more players are added. Nor is it a viable option for a dynamic gaming environment, like a MUD, or one that requires any kind of central administration or control.

In a client-server architecture, one node (the server) acts as a central intermediary between each player and all the other players, as well as between each player and the gaming environment. This simplifies client-side behavior and interaction with the network by shifting most of the game processing and networking demands onto the server. Provided the server is correctly configured and well monitored, this architecture offers the best approach for maximizing game performance as well as for updating the game environment to keep it fresh and challenging. But if the server fails or becomes overloaded, all players suffer the consequences. This is the most commonly used architecture for MUDs.

A server-network architecture combines the best elements of the other two architectures. Servers are connected in a type of peer-to-peer architecture. They continuously and directly share information about the state of the game and the players. Each server supports some subset of clients, which may be determined by network proximity or server load or any number of other factors. This approach has many advantages. First, it scales well: as more players enter the game, more servers can be added or the load can be altered to distribute the resources more evenly. Additionally, if one server exhibits network problems or has some hardware or operating system failure, another server can take over its load. This architecture is more costly to build and adds to the complexity of the network, but it offers the greatest control and flexibility for maximizing network performance and accommodating a more complex game playing environment.

# Virtual Reality: Living and Working in Cyberspace

As evidenced by the different types of chat environments and interactive games available on the Internet, cyberspace already functions as a popular and diverse virtual meeting place: a computer-simulated and computer-contained space through which people can navigate and in which people can interact. Through the Internet cyberspace extends into the workplace, the home, community centers, libraries, and elsewhere. But, in and of itself, cyberspace is amorphous; it lacks shape, definition, characteristics, and traits. It needs to be molded and given instructions as to what specific form it should take (e.g., a chat room or a forest filled with furry, anthropomorphic creatures).

There is no reason, therefore, why cyberspace cannot just as easily take on the form and functions of a workplace, complete with, for example, a reception area, individual offices, conference rooms, and a cafeteria. In addition to talking, playing, and socializing in cyberspace, one also might then work in cyberspace. In most business environments, communication and the sharing of information with colleagues and customers are integral parts of everyone's job, meetings are common and frequent occurrences, and networked computers are required equipment for performing one's job. Consider these the basic ingredients in creating any type of cyberspace workplace. It's not difficult to speculate on what form this virtual workplace might take, what effect it might have on how business is conducted and, more generally, how it might impact the way in which people communicate and interact.

The Internet brought email into the business world; and email quickly established itself as a commonplace and essential communication tool. Email offers a simple and effective means to do such things as share ideas, resolve problems, track issues, and assign work, either one-on-one or as part of group-based communications. Businesses that have incorporated email into their operation quickly come to rely on it. But email is asynchronous: the prompt delivery of a message is nearly guaranteed, but there is no way to know when its contents will be read. What's more, email is primarily text or document bound; it is fundamentally about the written word. In other words, email has its limitations.

The integration of chat programs into business environments has recently become more common. They make possible real-time message exchanges. They can also be an effective way to conduct informal meetings or brain-storming sessions, and to interrupt someone for an urgent matter without necessarily disrupting some other activity in which they may currently be engaged. Chat programs also work well as real-time support mechanisms, allowing customers to interact directly with technical staff, for example, in order to work a problem through to its resolution. But chat provides little structure and few control mechanisms. More often than not it is used for personal communications, or for asides and tangential observations. For instance, some people attending a meeting (in person or remotely) may meet simultaneously in a chat room to make observations on the meeting that they might not want to share openly. Consider chat a poor man's cyberspace environment.

MUDs, on the other hand, can take a business environment and recreate it in cyberspace. MUDs offer an environment that incorporates all the features of email and chat while further enhancing one's ability to communicate and interact with other employees; and a MUD can be as broadly or as precisely defined as seems appropriate for the particular needs of a business and its employees. MUDs achieve this by taking the distributed nature of the Internet and building on top of it a whole new paradigm of a work environment — a virtual office constructed in cyberspace with as many floors, rooms, desks, fax machines, and coffee makers as wanted. In doing so, each employee's computer suddenly becomes far more than a basic node on the network storing information, supplying productivity tools, and facilitating communication. It becomes an office within a larger environment of virtual offices (and possibly virtual buildings or campuses) and a portal through which one can communicate, attend meetings, deposit and retrieve information, collaborate on projects, and socialize. As such, MUDs can take personal interaction through the Internet to a level that email and chat can only approximate.

In general, MUDs provide business environments with the following features:

- An interactive, real-time work environment: when one person says something, everyone hears it immediately.

- A workspace free from the restrictions of walls and floors and buildings and geographic location: a MUD's client-server architecture allows everyone to work together in the same virtual office, or floor, or building, or campus; any type of space with any number of features can be created.

- Room to accommodate as many people as the situation dictates: virtual space has no size restrictions (and everyone can have an office with a window).

- An environment that adapts to the needs of people and the demands of work: MUDs are extensible; coffee machines, blackboards, rooms, whatever is needed, can be added as people see fit.

- Total control over who can, and who cannot, inhabit the space: MUDs are not anonymous places; everyone has their own unique account and cyberspace representation (i.e., avatar), which keeps the environment secure for its intended use.

- An automatic record of events and conversations: MUDs include a mechanism for recording what transpires, allowing missed conversations to be reviewed and permanent archives to be created.

MUDs can be created that mimic (through the definition of information objects) the specific offices, meeting rooms, cafeterias, copy rooms, and water coolers of existing environments. Or MUDs can just as easily be designed as abstractions, spatial metaphors that provide more playful but productive environments in which people can interact. Instead of an office building, an outdoor environment could be constructed, complete with forested areas and glades, a brook and a swimming hole, a picnic area, an ancient, expansive oak tree under which group meetings could be held, and a small log cabin where more private meetings could take place. Such an environment might even include forest animals, with squirrels to carry personal messages between employees.

Since a MUD is a context for problem solving, role playing, and general interaction with others, individuals can use movements within the virtual environment to capture and broadcast their movements in the physical world. By going to a virtual office and closing its virtual door, for instance, an individual can show others where they can be found. The same action may additionally indicate that this person is working privately on some issue and does not therefore want to be disturbed. In a virtual workplace one person's physical location may be a thousand miles from that of everyone else, but the geographic separation in no way diminishes that person's ability to indicate whether or not they are busy, if they are away from their desk, or if they have left for the day.

Another powerful component of virtual workplace environments in cyberspace is that they themselves can contain virtual tools. The use of such tools can potentially extend the whole topology of the Internet inside the virtual environment. For instance, a Web browser could be included as part of the virtual office. One person might bring up a page on the virtual browser related to an ongoing discussion and share it with a particular person or with an entire group. They might leave the page on a virtual tablet in a meeting room or on a table by the water cooler for people to find and read at any time. Others might jot notes on the tablet, commenting on the information; or they might simply leave their initials to indicate that they have read what was there. A common room or foyer area in the virtual office might include a virtual bulletin board on which group or company information could be posted for everyone to read. This virtual bulletin board could even be linked to a bulletin board in the physical world; and the information on one could be kept in sync with the other. In this way, the virtual and physical environments become extensions of each other, simplifying and enhancing the ability to communicate information within the company.

Another use of virtual environments in business is as a tool for coordinating the members of a group and their activities. This application can be especially useful for groups that provide support functions, in which individuals are often expected to handle a variety of emerging problems along with some number of recurring tasks. A virtual environment can readily identify what everyone is doing at any point in time, enabling it to function as a sort of real-time group management application. Work can be quickly assigned and more easily re-prioritized as conditions change, since the environment is acting to collect and make available information on the assignment, progress, and completion of tasks in a more timely and efficient manner than might otherwise be possible. A virtual environment can also make clear why someone might not be reachable, by allowing others to see the individual in question is driving home or is away from his or her desk on personal business. A virtual environment can also assist in broadcasting urgent, and less than urgent, messages. It might be used, for example, to announce that one of the computer systems is being taken down or that people are gathering for a lunch outing. All of these virtual environment capabilities enable people to remain more in touch with their co-workers and with events taking place at work, enhance teamwork and communication, and allow one's physical location to become less of a factor in how much and how significantly one can contribute in the performance of one's job.

The creators of virtual environments — whether designing those environments for work, play, or socializing — aspire to realize one of the highest goals of networked computing: to bring people closer together by fostering communication and enabling individuals to share information, resources, and themselves in new and unexpected ways. Virtual environments enable people to communicate in real time, individually or in groups. They also enable people to coordinate events, collaborate on their work, and participate in shared activities. The Internet has extended cyberspace across the globe, enabling people to gather and interact unconstrained by physical or geographic boundaries. Virtual environments like MUDs, and even simpler virtual constructs like chat rooms, beckon individuals to take themselves onto the Internet. They endeavor to transform the Internet from a communication system and information warehouse into a shared, global space free of walls, borders, and other types of barriers. These virtual environments represent a new and very powerful paradigm for computer use, one that promises to take the network closer to becoming the central feature in our computing environments and more of a feature in the routine performance of our jobs.

Some people hold up cyberspace and its virtual environments as examples of everything that is wrong with our increasing reliance on computers and our increasing use of the Internet. They consider cyberspace as something that promotes isolation and diminishes the role of face-to-face interactions in our lives. Ironically, when computers first arrived on people's desks at work and at home, they brought with them the promise to reduce the amount of time and effort it took to record information, analyze data, produce reports, and so on. They promised, in other words, to help us complete many routine and typically laborious tasks faster, and thereby free up more of our time for other pursuits. Yet there is no denying that more people are spending more and more time in front of computers; and this activity in itself is physically isolating. Instead of promoting isolation, however, the Internet and cyberspace may be supplying the very thing needed to pierce that isolation and enable people to stay in touch, feel connected, and participate in group activities. This is the promise of the Internet as meeting place.

# Strategies for Success: The Technology and Management of Interoperability

# 8

## Strategies for Success

By anyone's measure, the Internet ranks as one of the greatest success stories of the twentieth century. It is, moreover, a success story that no one individual or group can take credit for creating. Many people — from different disciplines, with different backgrounds and interests, and associated with different organizations (or with no organization) — contributed to the Internet's success. The fact that the Internet represents the combined efforts of so many individuals is yet another aspect of the Internet that sets it apart, and itself constitutes one of the factors behind its success. The contributions of these individuals took the form of innovative ideas, forward-looking engineering specifications, new types of computer software and hardware, and, most importantly, a passionate belief that what they were creating was far greater than the simple sum of its parts. How *much* greater no one anticipated.

The Internet's humble and obscure beginnings, which trace back to the earliest days of computer networking and the building of the ARPANET, offer no hint of the nature or scope of its future success or the extent of its future impact on the lives of so many people. Even after twenty years of continuous use (from the early 1970s to the early 1990s) among a growing number of academics, researchers, and others, the Internet managed to remain predominantly hidden from public view. Of the relatively few people who had access to the Internet during those years, fewer still recognized its potential to impart such sweeping and pervasive change. The Internet in its existing form seemed an unlikely subject for recurring news stories or for the creation of publications that focused on its technology, services, and impact; and it seemed equally incapable of becoming a commodity for use by the general public and taking on such forms as a wanted and expected service in public libraries and a popular place for the selling and reselling of merchandise.

Yet shortly after its privatization and commercialization, the Internet began its unyielding advance into the daily lives of millions of people. Without warning, it quickly and inextricably wove itself into both the workings of the business world and the encompassing fabric of society. For all intents and purposes, the Internet's transformation took on the appearance of an overnight sensation. But the factors that enabled this transformation and were, moreover, ultimately responsible for the Internet's success were actually many years in the making.

Evidence of the Internet's success — in the advertising of Internet addresses and the popularity and familiarity of Internet services, such as email and the Web — is all around us. But what enabled the Internet to grow and develop into such a commonplace and influential presence in our lives in so short a period of time? To which specific factors or features can we attribute its exponential and unchallenged rise in popularity and its ability to embed itself into so many different environments (e.g., the workplace, home, libraries), devices (e.g., mainframe computers, personal computers, handheld devices, telephones), and activities (e.g., buying/selling/trading products, school homework, business transactions and customer service, socializing)?

Technology played a leading role in the story of the Internet's success. Advances in technology, as exemplified by the development of packet switching and TCP/IP, brought about the Internet's creation, contributed considerably to its evolution, and continue to account for the speed and efficiency of its operation today. But the Internet's story is unlike earlier success stories (and revolutions) that can be attributed to some singular technological breakthrough, such as the Protestant Reformation and Gutenberg's invention of moveable type in the fifteenth century or the Industrial Revolution and Watt's invention of the modern steam engine in the eighteenth century. The engineering and implementation of the Internet's technology only goes so far in explaining the reasons behind its success. Another factor, which has surprisingly little to do with the mechanics of networking or the inner workings of computers and programming languages, played an equally compelling, although far more subtle, role.

This factor was one of strategy, and it includes the role played by individuals. In general terms, the Internet's success can be attributed to the sound planning and forward-looking work of the various organizations that acted as oversight bodies directing its management and development, such as the Advanced Research Projects Agency (ARPA), the Defense Communications Agency (DCA), the National Science Foundation (NSF), and the Internet Society (ISOC). Members of grassroots Internet groups, as well as innovative and generous individuals, also made considerable contributions that affected the Internet's development and kept the Internet free from the controlling influence of any one authority or group. In more specific terms — and often overlooked as reasons — the plain and deceptively simple strategies for achieving interoperability on the Internet, which were promoted and scrupulously adhered to by these organizations and individuals, may ultimately prove most responsible for the Internet's success.

These strategies of inclusivity and impartiality impacted both the development of the Internet's technology and the Internet's overall management. They were instrumental in building the Internet into the global, pervasive, and unrivaled force for change that exists today. That's because these strategies are responsible for no one owning the Internet and no one controlling what the Internet consists of, who can access it, what services it can offer, and what information it can (and cannot) contain. These strategies are also what's responsible for enabling any type of computing device to connect to the Internet and any type of information to travel across it.

The Internet continues to thrive and evolve today because of these same strategies for success. They are responsible for ensuring that the Internet remains operational, reliable, fast, efficient, adaptable, and equally accessible and empowering to everyone. This chapter considers what forms these strategies have taken, first from the perspective of the Internet's technology and then from the perspective of its management.

# The Internet's Technology: Conventions, Consensus, and Working Code

Interoperability is the hallmark of the Internet. As described in earlier chapters, achieving interoperability (i.e., enabling all types of devices to communicate and exchange information across the Internet despite differences in their composition and operation) was above all else a pursuit to be inclusive — to find and promote common ground in all things related to the Internet's technology. The strategies behind pursuing interoperability can be characterized by the creation of conventions, the obtaining of consensus, and the adoption of working code.

Consider the creation of conventions as the foundation of the Internet's technology upon which interoperability was implemented. Protocol specifications, issued as documents called RFCs (Request For Comments), emerged early as the most apparent and detailed representation of Internet conventions. These specifications quickly evolved into an accepted standard for documenting all Internet conventions, and they now form a large repository of information that defines the existing interoperability of the Internet, documents more than three decades of its history, and describes ongoing work to prepare the Internet for tomorrow's technology. The technological conventions contained in these specifications are not unlike linguistic conventions: the letters that make up a language's alphabet, the words built from those letters that fill a dictionary, and the syntax and standards of accepted usage that together make communication possible and that enable the exchange and storage of information.

Conventions defined and employed for use with the Internet allow Internet engineers and programmers, for instance, to work with a common definition for a packet of data and to talk about the different functions and features of a network through commonly understood terminology and abstractions (e.g., dividing the operations of a network into a series of stacked layers). These same conventions perform much the same function for the devices attached to the Internet. They enable these devices to be programmed to recognize common definitions and to behave in an orderly, prescribed manner, thereby making intercommunication and the exchange of data possible.

Consider the pursuit of consensus as the path upon which interoperability was reached. Government organizations, large and small corporations, grassroots groups, and individuals voiced different wants and needs with respect to the Internet's development and its technology. Consensus among all the interested parties translated into a commitment to fundamental interoperability, since only through arriving at consensus could everyone possibly obtain something of what they wanted or needed without causing the Internet as a whole to fracture into any number of separate and isolated networks. Consensus made it possible for networks to remain independent and different, but still interoperate at some baseline level and allow traffic to flow from one network to the next across the Internet.

The Internet's protocol specifications and the conventions they contain manifest both the pursuit and attainment of consensus. The place in the specifications where this is most visible (i.e., in its most practical and effective form) is in the categorization of protocol requirements as required, recommended, or optional, and the associated categorization of protocol implementations as compliant, unconditionally compliant, and conditionally compliant. Only through consensus could some form of compliance be reached; and without broadbased compliance to networking protocols, achieving any measure of interoperability on the Internet would have been impossible.

The creation of conventions, therefore, allowed everyone to speak the same language and lessened the difficulty of comprehending and employing the abstract constructs of computer systems and networking. Meanwhile, the pursuit of consensus allowed everyone to arrive at compromises that helped meet the needs of the greatest number of people while recognizing, and attempting to accommodate or ameliorate, differences that might otherwise obstruct interoperability.

But working code is where interoperability happens. Consider the adoption of working code as the catalyst that first brought interoperability (and networking itself) into existence and that continues to provide the intelligence through which the Internet functions. The Internet had to start somewhere, and it had to be able to expand and evolve as time passed, as more networks were created and wanted to interconnect, and as ever increasing numbers and types of devices demanded access. Working code, therefore, transformed the Internet from networking theories, written specifications, and an assortment of experimental networks into a functional, dependable, fast, and efficient global network for transporting packets of data, sending and receiving email, surfing the Web, playing games, interacting in online communities, and so much more.

Defining and implementing interoperability on the Internet was necessary if the Internet was to succeed. Many different factors contributed to fulfilling this requirement. In terms of the Internet's technology, these factors took the form of new types of computing devices, new approaches to computer programming, new models and concepts of computer engineering, new ways to design computer systems and their components, and new thinking on how computers could be used. Several of the factors that were key to implementing the Internet's interoperability are described below, first briefly in a list to provide some sense of the nature and breadth of this ground-breaking work, and then at greater length in individual sections.

- **Resource sharing**: the name given to the original, motivating principle behind the creation of computer networking (in general) and the pursuit of interoperability (in particular).

- **Interprocess communication**: the name given to the method through which a program on one computer interacts and communicates with a program on another computer, establishing how and where interoperability takes place.

- **Logical connections**: the name given to abstractions that enable physical resources (e.g., a tape drive or CD-ROM unit) to be identified and referenced by logical or generic names, introducing a measure of modularity that simplifies the sharing and interoperability of networked resources.

- **Layering**: the name given to a framework that simplifies discussing, engineering, and implementing interoperability by dividing network functions into a series of stacked layers.

- **Virtualization**: the name given to abstractions that facilitate interoperability through the building of virtual components (i.e., conceptual, information-based components as opposed to physical, hardware-based components) that are designed free of dependencies on any specific type of hardware or computer architecture.

- **Efficiency and equity in protocol specifications**: the names given to design objectives that contributed to interoperability by keeping protocol specifications as simple and as architecture-independent (i.e., not favoring any one type of computer platform) as possible.

- **TCP/IP**: the name given to the protocol suite that implements the essential, underlying interoperability of the network.

- **Gateways**: the name given to the devices that interconnect networks and facilitate interoperability by localizing and accommodating any differences between dissimilar networks.

Each of the items presented in the list above are described in more detail in the sections below.

## Resource Sharing

Resource sharing was a fundamental goal of the original ARPANET engineers and it remains today one of the driving forces behind the creation and use of most networks. Even the simplest network (e.g., a home computer network with a couple of personal computers sharing a printer and a single Internet connection) illustrates the basic, inherent value of resource sharing over a network. When resource sharing is implemented on a larger scale, such as in a typical business or academic environment, the value of resource sharing becomes that much more pronounced.

At the time of the ARPANET, resource sharing focused on remote computer access and file sharing. Computer resources were scarce and expensive; they consisted of large pieces of equipment that were housed in isolated, access-restricted locations. Resource sharing by means of the network established remote access to those resources, allowing them to be used by more people and in the performance of more work. The focus of resource sharing, however, widened over time, as the resources themselves changed, new types of resources became available, and both how and why networked resources were used altered and grew. Computer equipment became small, portable, and inexpensive; and the use of computers became an everyday activity of millions of people. Accordingly, resource sharing became a way to reduce the gross duplication of equipment that had become commonplace in most networked computing environments and, in the process, reduce the excess cost and additional service overhead that this duplication entails. In a business environment, for example, resource sharing means one networked printer for a group of people in place of any number of separate, dedicated printers attached to each computer, and one installation of an application on a centrally located server in place of any number of separate installations on any number of individual computers.

Then and now, resource sharing required two conditions to be satisfied before it could be realized. First, it needed a network to physically interconnect the resources to be shared. Second, it needed some measure of interoperability to make the process of accessing and using those resources feasible from any host on the network. On the ARPANET and early Internet — just as on today's Internet — resource sharing had to take place among heterogeneous devices (i.e., among computers and other equipment running different operating systems, supplied by different vendors, or exhibiting any number of other fundamental incompatibilities). This requirement demanded a more far-reaching approach to resource sharing than any single piece of technology or any one engineering solution could possibly offer. While it's true that specific issues of incompatibility between any two devices could be resolved by translating the functions of one system so that they would be understood by the other (much like translating a page of instructions from French to German), this was not, however, the path to interoperability. What was needed was a more generalized (i.e., more abstract) approach that would not only allow existing incompatibility issues to be resolved for *any* two devices that needed to talk to each other and interact, but would also accommodate the incompatibilities of hardware, software, and operating systems that had not yet been built.

One approach taken in the effort to implement generalized interoperability in resource sharing was the specification of protocols. Protocols were created to define precisely how host computers and other networked devices would communicate with the network and with each other. The first two application-layer protocols — TELNET and FTP — made resource sharing possible by defining a common language and a framework that precisely controlled how each networked computer communicated and exchanged data with any other computer on the network. These protocols defined the very rules of interoperability for remote computer access and file sharing, enabling the ARPANET engineers to realize one of their principal objectives in building the first packet-switched network. As the network grew and its use expanded, subsequent protocols defined and implemented different types of resource sharing, such as SMTP for transferring email and HTTP for transferring Web pages. These protocols, like those that came earlier, measured their success by how well they managed to achieve interoperability.

Prior to the ARPANET's creation, resource sharing established a goal that served to justify and motivate the building of the first network. It held forth the promise of more economical and egalitarian use of scarce and expensive computer resources, provided interoperability could be attained. Today, resource sharing gives networking a purpose that can be measured in time, money, and ultimately, in protocols. It makes computer systems, the information they contain, and the devices they control accessible to virtually anyone in any location using any type of computing device. Resource sharing is, therefore, an integral component of the Internet. Its contribution to the Internet's success can be found in every service the Internet offers. But what makes resource sharing possible is the underlying interoperability delivered by protocols. Without broad-based interoperability, neither the Web nor email would have reached beyond the academic or scientific communities from which they began, and most home and business computers would be restricted to the resources, services, and information in their local and isolated environments.

## Interprocess Communication

Computers operate through the execution of programs. Each time a program executes it is performing a task, or some sequence of tasks, such as reading the contents of a file for display on a monitor or writing out the contents of a new or modified file to a disk drive. A process is a general term used to describe this fundamental task-based program execution.

Since resource sharing consists of one computer making one or more of its resources available to another computer, the two computers require some way to communicate and exchange information for resource sharing to take place. This communication occurs through interconnecting processes on the two computers across the network, thereby connecting the two computers at the level of their processes, and is named interprocess communication. Interprocess communication does not, however, only take place between computers across a network; it also occurs between processes on a single computer. In general, interprocess communication allows one application to control another application, and it allows several applications to share the same data without interfering with one another.

The protocols that were created to make resource sharing possible relied on interprocess communication to provide the connection point for any two computers to exchange any type of information. This approach was sufficiently generalized to fulfill the needs of the first network services — TELNET and FTP — and to make possible all subsequent Internet host-to-host services. When two computers are negotiating a connection — shaking hands and identifying themselves and their capabilities — they do so through interconnecting processes; this act effectively forms a bridge across the network that enables them to send and receive information. Each and every task computers need to perform across the network requires this same, basic interprocess communication. Interprocess communication is, therefore, both how and where interoperability occurs from the perspective of the individual networked tasks that computers routinely perform.

## Logical Connections

In order for resources to be shared, they first need to be named. Logical connections are names for networked resources, and they serve to absorb specific hardware, software, and architectural differences between two computers as part of the overall effort to facilitate general interoperability and resource sharing. Interprocess communication provides the physical connection points (i.e., in technical terms, the channels, routes, paths, ports, and sockets) that establish the link between two host computers. Logical connections provide an abstract designation that make it possible to reference these connections, and the resources they compose, in a simple and modular way.

A logical connection designates a resource so that it can be easily referenced and used without knowing anything about its physical characteristics. It allows, for example, a remote computer to access a local computer's tape drive without needing to know about such details as how the tape drive is connected, the type of tape it uses, or how it is controlled. The local computer handles the specifics of the interaction with the physical device, while the logical connection provides an abstract view of the resource for it to be seen and used across the network. In this way, a logical connection allows a resource name to remain the same even if the actual resource it references changes.

For example, a network could include a logical connection named *tape-drive* as a shared resource for backing up files from any of a dozen computers. Because the logical connection is simply a name, it can be configured to reference any one of several possible devices. The specific device used might change from one day to the next due to maintenance, an upgrade, or other issues. Or it might reference a room filled with tape drives and transparently call upon some program to select the first one that becomes available. Or it might reference a virtual tape drive that creates a tape archive file on a computer center disk drive rather than copying the files to tape. An individual seeking to access the resource through a logical connection could use it without knowing any of the details related to its location, configuration, or properties. Moreover, none of the details related to the physical characteristics of the resource need to be known to the individual's computer either.

Logical connections, therefore, enhance and simplify interoperability by introducing a measure of modularity. Specific components — individual resources and even hosts — are free to come and go on the network and change, while the references to those components, whether used directly by individuals or incorporated into programs, can remain unaffected and still access the same general resources.

## Layering

Layering is a framework for describing the different, but interconnected elements of networking (e.g., the 7-layer OSI network model). Like logical connections, layering is an abstraction. It contributes to interoperability by dividing large, complex tasks and processes, like the process of transmitting data from one computer to another across a network, into discrete, modular components.

Although ARPA and the International Standards Organization (ISO) presented different conceptual views of the layers that networking could be divided into, both organizations treat layers as functional groupings arranged in a stacked hierarchy. The following general description applies to the views presented by both organizations. The first, or lowest, network layer represents the most concrete and physical area of networking; it deals with the incoming and outgoing transmission of electrical signals as data passes between a computer's networking hardware (e.g., an ethernet card) and the connected network. Each layer becomes progressively more abstract as the data is converted from electrical signals into the zeroes and ones of a computer's digital language, and as each successive layer accepts the data, handles it according to its specific responsibilities (e.g., packaging it into IP packets), and hands it off to the adjoining layer. The last, or topmost, network layer represents the most abstract area of networking; it deals with the applications that use the network (e.g., FTP, email, a Web browser) by capturing and preparing outgoing data for transmission across the network and, conversely, by accepting incoming data for its display or use in the associated application.

This division of networking into separate and distinct layers makes it easier to discuss and diagram all the various operations that need to occur for two computers to communicate. Moreover, it greatly simplifies the creation of individual protocols to handle each particular operation. Each layer in the hierarchy operates independently on the data, with its own particular tasks to perform. Accordingly, each layer has one or more protocols to handle these tasks, such as TCP in the transport layer to check and ensure the data's reliability and IP in the network layer to handle packaging the data into packets. Data passes down the stack from the application to the transport layer in order to be sent across the network; and as it does a protocol in each layer accepts, handles, and passes on the data according to its particular definitions. Data passes up the stack on the receiving end, going through the process in reverse, until the original application-layer data has been reconstituted on the remote machine.

The interaction between layers is kept to a bare minimum — only what is needed to allow one layer to hand off the data to the next. The independence of the layers, like the overall framework of layering, assists interoperability by dividing network operations into discrete and independent components. This means that new network operations (and new protocols) can be introduced without compromising existing functionality, and existing operations (and protocols) can adapt without necessitating any changes in the operations performed in adjoining layers.

## Virtualization

Another method of abstraction that serves to bridge the differences between computer systems and facilitate interoperability is a process called virtualization. Virtualization involves taking a particular piece of hardware, such as a computer terminal or a tape drive, and creating a virtual representation of the device (i.e., a model of the device made exclusively from information). The information describes the device in terms of its features and functions, and defines precisely what it can and cannot do; the information also comprises programming logic that controls the behavior of the virtual device and implements how it is accessed and used. In this way, virtualization provides a way to create generic hardware components: common information-based representations of devices and objects distinct from any one computer's physical implementation of such a component.

A good example of virtualization — and one of the earliest — was the creation of the Network Virtual Terminal (NVT) in the TELNET protocol specification. The NVT accommodated differences among the terminal types produced by various manufacturers by providing an abstract model of a terminal. Like virtual memory, which supplied a resource that acted as memory from the perspective of the operating system without the physical counterpart, the NVT supplied a description of a generic display and keyboard that interacted with the local operating system like any other terminal but without the physical counterpart.

The virtual terminal abstraction comprised by the NVT provided an elegant and efficient application-level solution to the difficult problem of enabling any type of computer terminal to work with any type of networked computer. The NVT delivered this essential form of interoperability (neither TELNET nor FTP would have been possible without it) by representing a generic terminal type that any two interconnected computers could recognize and use in the exchange of information. Without the NVT, it would have been necessary for each host computer on the network to recognize hundreds or thousands of different terminal types — precisely the circumstance that the Internet's engineers were working to avoid.

Virtualization, as exemplified by the NVT, is a powerful mechanism for building any type of virtual device. It can be used to create generic versions of existing devices suitable for absorbing hardware differences, such as the NVT. It can also be used to create new types of devices, like a tape drive that archives files for safe keeping but does not necessarily copy those files to tape (i.e., it might store the files on a CD or on a hard disk). It can even be used to create representations of entire computer systems and networks, which might be useful in conducting simulations of newly designed computer components or in researching new networking theories. With respect to the Internet, virtualization is a key factor in the effort to achieve interoperability. It provides a flexible and powerful mechanism for both overcoming existing hardware differences and accommodating future differences as new types of computers and networked devices connect to the Internet.

## Efficiency and Equity in Protocol Specifications

Writing a protocol specification is a difficult balancing act. The objective is to define the operations that constitute how two computers can interact and exchange information in the performance of some specific task, such as the forwarding of a mail message or the display of a Web page. The problem is that the specification must be sufficiently generalized to allow for a wide variety of implementations on any number of computer platforms. But it must also find a way to remain sensitive to existing differences among those platforms in defining operations that might favor or hurt one or more potential implementations. In the end, a protocol's success is measured by how well its specification facilitates interoperability.

Two guiding principles serve to promote interoperability in the writing of a protocol specification. The first principle is efficiency. For a protocol to be considered efficient, its operations need to be kept as simple as possible and preferably few in number. The greater a protocol's efficiency, the less of an impact it will have on a computer's overall performance and its available resources. The second principle is equity. Equity is a matter of fairness, and it demands that protocol operations be defined in a manner that is not biased towards any existing computer manufacturer or architecture. Equity is also a matter of accommodation, and it demands making allowances for existing conditions that might otherwise keep one or more parties from being able to implement the protocol on a specific computer platform.

The pursuit of efficiency and equity in protocol specifications, therefore, focuses the attention of those who participate in creating these specifications on the overall goal of interoperability. Interoperability means that any one compliant implementation of a protocol specification will be able to interact correctly with any other compliant implementation of the same protocol, whether the respective computer systems on which the implementations reside are the same or different. There exists, therefore, a delicate but essential balance in any protocol specification between the level of specificity that will ensure this interoperability and the generality necessary to leave implementation details to the people engineering and building the conforming products. How well this balance is achieved often affects how widely a protocol is accepted among the diverse group of individuals, groups, and companies that provide the products in which the protocols are implemented.

The means to achieving a reasonable level of efficiency and equity in any protocol specification relies on the help and input of a diverse group of interested individuals and experts. Any one requirement in a specification might be easy to implement on one computer platform while nearly impossible to implement on another. Individual contributors involved with the writing of a specification — who often represent specific companies with existing products — can use their participation to request changes to requirements that might hinder their ability to implement the protocol. It follows that arriving at compromises is a typical occurrence in the writing of a protocol specification.

An early debate over email provides a good example of both conflict and compromise. The File Transfer Protocol (FTP) formally incorporated email into its specification in 1973 by defining both an independent mail command and the special characteristics of a mail file. Email represented a distinctly different and highly specific type of file transfer; and many people anticipated that FTP use would increase significantly due to email, and most of the file transfers would be automated through email programs rather than done manually by individuals. But an existing FTP requirement stating that a user had to identify himself or herself before being allowed to perform any operation on an FTP server presented a problem when it came to using FTP for transferring email. The requirement meant that whenever a user (or a program running on a user's behalf) connected to any FTP server for any reason, the FTP server first had to issue prompts for a login name and a password to be entered. Some people complained that this requirement was simply unnecessary when applied to email transfers, while others feared that requiring authentication through a login and password would lead to charging mechanisms for delivering email.

Efforts were made to change the FTP specification to make the authentication operation optional rather than required. But the purpose of this requirement was to ensure consistency among FTP implementations (i.e., effect interoperability) and to provide those who needed it a means to authenticate remote users. Certain operating systems required authentication, either for accounting purposes or some other reason; and those that did not require authentication, it was reasoned, would not be negatively affected by being made to use it. The FTP specification did not, however, require that the login and password supplied by the user be used to authenticate that user (i.e., check that a matching account existed on the local computer); it only required that the user be prompted and that something be entered. The authentication process — as opposed to the operation of prompting for a login and password — was considered an implementation detail outside the bounds of the FTP specification, which meant it could be handled any way that suited the needs of those writing the particular FTP implementation. This distinction enabled a compromise to be reached that would fulfill everyone's needs. The operation that prompted for a login and password was retained as a requirement, but a generic login name and password were added to the FTP specification to effectively allow anonymous authentication. This defeated the possibility of using such information as part of a charging mechanism.

## TCP/IP

In order for two computers to exchange data — whether the data is part of an email message or part of a Web page — they first need a common, predefined language and framework through which to communicate. In order for information to travel across the Internet — to leave one computer and then hop from computer to computer and network to network until it reaches its destination — it first needs to be in a common, predefined format. The suite of networking protocols named TCP/IP furnishes the Internet with the common language (i.e., the lingua franca) that fulfills both of these network functions, and in doing so it implements the core network-level interoperability of the Internet.

TCP/IP defines (through protocol specifications) and implements (through protocol software installed on every host computer connected to the Internet) how data is reliably and efficiently controlled as it is transferred from one computer to another (TCP) and how data is uniformly packaged and quickly transmitted as it journeys across the Internet (IP). Its creation marked the beginning of the Internet, alongside the creation of gateways and DNS.

Before TCP/IP, the network — not the hosts — was responsible for ensuring the reliability of the data that traveled the network. The hosts presumed that the data they received arrived undamaged, complete, and with the packets correctly sequenced. But the interconnection of dissimilar networks required that the intelligence behind the network be moved to the hosts, and that the network be considered unreliable. The hosts would now need to reorder packets as they arrived, provide verification that packets were received, perform error checking by testing that each packet arrived undamaged and request the retransmission of lost or damaged packets, and control traffic flow by limiting the number of packets in transit. The creation of the Transmission Control Protocol (TCP) enabled each host to perform these essential networking tasks, providing the reliability and robustness previously supplied by the IMP subnet.

While TCP provides the necessary data control and validation functions, the Internet Protocol (IP) simply handles the packaging and moving of the data. The design of IP is purposely generic and simplistic: "an irreducible minimum for the functionality of anything one would be willing to call a network."[1] IP's minimalistic approach to the packaging of data into small and simple packets is precisely what enables data to travel quickly across the Internet and to pass safely between dissimilar networks.

TCP/IP defines and implements the core interoperability that is the Internet. It is more responsible for the Internet's success than any other single feature. Remove FTP, TELNET, or the World Wide Web, and the Internet is diminished. Remove TCP/IP, and the Internet is effectively erased. One of the principal reasons TCP/IP acquired such a dominant role in the Internet's development is that is makes few, if any, assumptions about the devices on which it resides. Nearly any device that has the capacity to tell time (which is needed to signal for the retransmission of lost or damaged packets) can include some implementation of TCP/IP. This means that TCP/IP can turn any existing (or future) electronic device into something capable of connecting to the Internet and exchanging information across the network. The same basic TCP/IP that enabled mainframe computers, workstations, desktop personal computers, and laptops to become part of the Internet can now be found on Internet-ready cellular telephones, personal digital assistants (PDAs), and other mobile, handheld devices. This facility of TCP/IP is what makes it the very definition of interoperability.

## Gateways

Gateways (or routers) isolate and account for the differences between interconnected networks by routing data from one network to the next and making any required changes to the format of the data packets. Gateways function as virtual hosts on the Internet. To the network, a gateway appears as any other type of host computer. It has a name, an IP address, and a network interface (e.g., an ethernet connection). Its purpose, however, is to function as a bridge, interconnecting two or more networks (or subnets).

A gateway is configured with information about each of the networks it interconnects. This information includes how to route outgoing traffic so that it can be delivered to its destination, or to the next relay point in its continuing journey across the Internet; and it includes how to direct incoming traffic to its intended destination. Configuring a gateway with this information eliminates the need to program the individual host computers with such routing information; they only need to know about the gateway's existence. Information stored in the gateway also includes details about the features and data packet formats of each of the networks it interconnects. This information allows the gateway, for example, to perform any operations on the data packets (e.g., adjusting the packets' header contents or modifying the packets' size) in order to meet the requirements of the adjoining network to which the packets are being routed. By absorbing the differences between two interconnected networks and localizing the information required for this work to a single device, gateways greatly reduce the complexity of the individual networks. Moreover, they implement a critical component of the Internet's fundamental interoperability, in particular as it relates to the Internet's role as a global, distributed packet-switched network.

Gateways have purposely been kept simple and focused; they do not, for instance, handle any protocols above IP. Their role on the Internet, originally and today, is to provide a single place to encapsulate network differences and to simplify the routing of traffic from one network to the next. In performing this role, gateways enable older networks to remain operational as newer technology is introduced, and they enable new technology and new types of networks to be used without introducing incompatibility issues or making demands on the other networks. As such, gateways extend the modularity, adaptability, and interoperability of the Internet to the individual networks that together make up the Internet. They accomplish this by enabling each network to operate independently while also enabling each network to interoperate with other networks, no matter how similar or dissimilar those other networks may be.

# The Internet's Management: Decentralized and Democratized Control

Technology defines the interoperability of the Internet. The success of this technology can be measured directly by how well computers interact with the network and with each other in the performance of standard tasks, such as sending and receiving email or clicking on a link to view a page on the World Wide Web. The conventions, consensus, and working code that contributed to the Internet's success by building interoperability into the technology represent the hard work of many individuals over the course of many years. Equally important, interoperability is also the product of the way in which the Internet has been managed, from its very beginnings as a U.S. government project run by ARPA, through subsequent control by other government agencies, and continuing today in its privatized control by the organizations described below.

The strategies developed and pursued by the various organizations that were responsible for the Internet prior to its privatization and commercialization demonstrated a remarkable penchant for soliciting help, building consensus, and inspiring individuals to donate their creations so that they could be used and adapted freely by others. The efforts of these organizations to be inclusive and to carefully balance their control over the technology's development against the openness with which the technology was defined, built, and made freely accessible to all were without precedent. More importantly, they imparted a fundamental and irreversible democratizing influence that carries through to the Internet's management today and that continues to inspire the creation of an ever growing number of new Internet-based products and services.

The management of the Internet is currently conducted by a collection of organizations described in the following sections. These organizations provide the policies and framework for individuals, as well as corporate and government representatives, to meet, discuss their common ground, their differences, their needs and objectives, in the interests of maintaining interoperability across the Internet, resolving problems, and planning for the Internet's future. The division of labor follows along the lines of different perspectives regarding the technology, with some groups focused exclusively on the details of the technology, others focused on the larger, architectural issues, and still others managing the process itself of documenting and establishing the evolving standards for interoperability. The overall effect is a decentralized management structure, with groups of individuals working separately on different pieces and aspects of the technology. These groups produce specifications in the form of RFCs as part of the process of documenting their work and making that work freely available.

## The Internet Society

The Internet Society (ISOC) is an open membership organization formed in 1992 that promotes global coordination and cooperation with respect to the maintenance, development, and availability of the Internet and its technology. It consists of representatives from over 180 countries, including more than 16,000 individual members and more than 150 corporations, non-profit and trade associations, foundations, educational institutions, and government agencies. These members share an interest in the existing operation of and future plans for the Internet. The ISOC provides leadership in making decisions related to Internet policy, standards, and legislation.

Currently, most work on Internet standards is performed under the auspices of the ISOC, as explained in the following sections. However, as the Internet continues to grow, more and more organizations are choosing to develop their own Internet standards. Examples of such organizations include the International Telecommunications Union Telecommunications standards group (ITU-T), the International Institute of Electrical and Electronic Engineers local area network standards group (IEEE 801), the Organization for Internationals Standards (ISO), the American National Standards Institute (ANSI), and the World Wide Web Consortium (W3C).

## The Internet Engineering Task Force

The Internet Engineering Task Force (IETF) is primarily responsible for the engineering and development of Internet technologies. Its membership consists of network designers, operators, vendors, and researchers who are concerned with the evolution of the Internet's architecture as well as its daily operation. The organization comprises more than 100 working groups, each focused on one or more specific issues and managed by an area director. The IETF was formally established by the IAB (see below) in 1986, although it had operated before that in a less structured environment. The IETF's operations are funded in large part by the ISOC, but it operates as an independent, self-managed group, with no corporation, board of directors, membership, or dues. It describes its mission as follows:

> Identifying, and proposing solutions to, pressing operational and technical problems in the Internet;
>
> Specifying the development or usage of protocols and the near term architecture to solve such technical problems for the Internet;
>
> Making recommendations to the Internet Engineering Steering Group (IESG) regarding the standardization of protocols and protocol usage in the Internet;
>
> Facilitating technology transfer from the Internet Research Task Force (IRTF) to the wider Internet community; and
>
> Providing a forum for the exchange of information within the Internet community between vendors, users, researchers, agency contractors, and network managers.[2]

## The Internet Research Task Force

The Internet Research Task Force (IRTF) is an organization of research groups focused on the long-term technical strategies and engineering relating to the Internet, specifically: protocols, applications, architecture, and technology. Members of the IRTF research groups are individual contributors, not company-sponsored members. The IRTF is supported in part by the IETF and the ISOC.

## The Internet Engineering Steering Group

The Internet Engineering Steering Group (IESG) provides technical management of the IETF's activities and the Internet standards process. As part of the ISOC, it functions as a formal standards body with responsibilities that include the final approval of specifications as Internet standards. The IESG ratifies and amends output — including RFCs — produced by working groups of the IETF. It also starts up and terminates working groups and ensures that draft documents submitted as RFCs by other groups are correct.

## The Internet Architecture Board

The Internet Architecture Board (IAB) is a technical advisory group of the ISOC. In this capacity, it is chiefly responsible for defining the overall architecture of the Internet and providing guidance and broad direction to the IETF and IRTF. But it also functions as the technology advisory group to the ISOC. Here its focus is on the larger, more general issues of the Internet, long-term planning, and coordination of the various ISOC-related organizations. Emerging Internet trends — popular new uses of the Internet that were not anticipated — are also a focus of the IAB.

## The Internet Assigned Numbers Authority

The Internet Assigned Numbers Authority (IANA) operates as the registrar for the IETF's technical specifications, as mandated by the IAB. For instance, as new file types are created for the latest technologies, the IETF will adapt or amend existing specifications to provide a standard definition (a name and/or a means of reference, such as a MIME type for email) that interested individuals or corporations can then use. This type of standardization is essential to ensuring the Internet's continued ease-of-use and to facilitating the creation of new applications for their use over the Internet. IANA keeps track of this type of technical information so that there is a single, reliable source for people to consult.

Prior to 1998 and the creation of ICANN (the Internet Corporation for Assigned Names and Numbers), a non-profit organization that manages the day-to-day functioning of DNS, IANA was also responsible for the root servers of DNS and IP allocation. Nowadays, ICANN provides funding for IANA and handles the more divisive, politically-charged technology issues of allocating and storing Internet names and addresses, while IANA serves simply as a repository of information consulted by engineers and programmers in the development of new protocols, applications, and services for the Internet.

# Milestones, Netiquette, and Jargon

## Computer History Milestones

The following listing of milestones in the history of computers is more representative than exhaustive.

1842     Charles Babbage designs an Analytical Engine to perform general calculations automatically.

Ada Byron King, also known as Lady Lovelace, becomes the first programmer.

1890     Herman Hollerith designs a system to record census data on punch cards, later commonly referred to as Hollerith cards. Data is stored as holes in the cards, which are interpreted by machines with electrical sensors. Hollerith starts a company that will eventually be merged with three other companies to later become IBM.

1939     The first electronic digital computer, the Atanasoff-Berry Computer, or ABC, is built by Dr. John Atanasoff along with his graduate student, Clifford Berry. His project was funded by a grant of $650.

1943     J. Presper Eckert, Jr. and Dr. John Mauchly design and build the first large-scale programmable electronic digital computer, the ENIAC, at the University of Pennsylvania. The ENIAC weighs 30 tons, contains 17,000 vacuum tubes, requires a 30-by-30 foot room, and cost $500,000 to build.

1945    John von Neumann proposes the logic design for the first stored-program computer, establishing the architecture that formed the basis for modern computers.

1951    Eckert and Mauchly build the first general purpose, electronic digital commercial computer, the UNIVAC, which correctly predicts that Dwight D. Eisenhower would win the presidential election after analyzing 5% of the tallied vote.

1952    Dr. Grace Hopper proposes the first high-level programming language, FLOW-MATIC, using symbolic notation rather than machine language.

1954    The first operating system is developed by Gene Amdahl and is used in the IBM 704.

1957    An IBM team, led by John Backus, designs the first commercial high-level programming language, FORTRAN (FORmula TRANslator). It is used for mathematical applications, including solving engineering and science problems.

1958    The first computer to use the transistor as a switching device, the IBM 7090, is introduced, marking the start of the second generation of computers.

1960    The COBOL (COmmon Business Oriented Language) programming language is created, using English-like phrases and aimed at business application development.

1964    The first computer to use integrated circuits (chips), the IBM 360, is announced. This smaller, faster computer marks the start of the third generation of computers.

1965    The CTSS (Compatible Time-Sharing System) operating system is introduced. It allows several users to share access to a single computer.

        The BASIC programming language is created; development is led by Dr. John Kemeny.

1970    The first version of the Unix operating system, developed by Ken Thompson and Dennis Ritchie at AT&T, is released on the DEC PDP-7.

          Fourth generation computers arrive with LSI (Large-Scale Integration) chips containing as many as 15,000 circuits.

1971    The Pascal programming language is introduced, developed by Nicklaus Wirth and engineered for teaching structured programming concepts.

1972    At AT&T, Dennis Ritchie develops the C programming language.

1973    Part of the Unix operating system is implemented in C.

1975    The first microcomputer, the Altair, is introduced.

          The first supercomputer, the Cray-1, is announced.

          Bill Gates and Paul Allen found Microsoft.

1976    Digital Equipment Corporation introduces its popular minicomputer, the DEC VAX 11/780.

1977    Steve Wozniak and Steve Jobs found Apple Computer.

1978    Dan Bricklin and Bob Frankston develop the first electronic spreadsheet, called VisiCalc, for the Apple computer.

1979    Bjarne Stroustrup of AT&T introduces "C with Classes."

1981    IBM introduces the IBM personal computer, called the PC AT. It includes the DOS operating system supplied by Microsoft.

1984    Apple introduces the Macintosh computer, incorporating the first widely available graphical interface using icons, windows, and a mouse device.

1985    C++ is created from Stroustrup's "C with Classes." It includes the concept of reusable software components called objects, introducing object-oriented programming as a new programming paradigm.

1988    Work on standardization of C++ begins.

1989    The American National Standards Institute (ANSI) publishes the first standard for the C programming language.

1990    Microsoft introduces the Windows operating system (Windows 3.0) for IBM personal computers, including a graphical user interface.

1995    Sun Microsystems announces Java, an object oriented programming language used to create Web pages with dynamic content, enterprise applications, and more.

1997    ANSI C++ is approved.

## ARPANET Milestones

The following milestone information focuses on the ARPANET. Information during the same period that relates more directly to the early Internet appears in the next section.

| ARPANET/Internet Growth | | | |
|---|---|---|---|
| Date | Hosts | Date | Hosts |
| 1969 | 4 | 10/85 | 1,961 |
| 04/71 | 23 | 02/86 | 2,308 |
| 06/74 | 62 | 11/86 | 5,089 |
| 03/77 | 111 | 12/87 | 28,174 |
| 08/81 | 213 | 07/88 | 33,000 |
| 05/82 | 235 | 10/88 | 56,000 |
| 08/83 | 562 | 01/89 | 80,000 |
| 10/84 | 1,024 | | |

1957    The Soviet Union successfully launches and orbits the Sputnik I artificial earth satellite, prompting widespread fears of technological inferiority in the U.S.

1958    The Advanced Research Projects Agency (ARPA) is created by the U.S. Department of Defense.

1960    In his paper, "Man Computer Symbiosis," J. C. R. Licklider describes connecting computers together by means of networked communication lines.

1961    Director of Defense Research and Engineering (DDR&E) assigns a Command and Control project to ARPA.

       Leonard Kleinrock publishes his paper, "Information Flow in Large Communication Nets," the first paper on the theory of packet switching.

1962    J. C. R. Licklider joins ARPA as director of its Information Processing Techniques Office (IPTO), which was established to coordinate ARPA's Command and Control research.

1963    Licklider forms distributed group of engineers and researchers and names them the Intergalactic Computer Network.

1964    Licklider leaves IPTO in July, succeeded first by Ivan Sutherland and then Robert Taylor.

       Paul Baran describes theory of packet-switched networks that are distributed and redundant in his paper, "On Distributed Computer Networks."

1965    ARPA sponsors study on a time-sharing, cooperative computer network. "The Experimental Network" is created through directly connected, dedicated 1200 baud phone lines between the TX-2 at MIT in Boston, the AN/FSQ-32 at System Development Corporation in Santa Monica, and the Digital Equipment Corporation computer at ARPA.

1966    Lawrence G. Roberts distributes the first formal ARPANET plan, "Towards a Cooperative Network of Time-Shared Computers."

1967    ACM Operating Systems Symposium in Gatlingberg, Tennessee. Initial plan for the ARPANET is unveiled. At the same conference, a paper by Davies, Bartlett, and Scantlebury presents the proposed National Physical Laboratories (NPL) packet-switched network for the United Kingdom.

1968     ARPA Program Plan No. 723, 3 June. "Resource Sharing Computer Networks" prepared by Lawrence G. Roberts. Both documents advocate importance of networks for computer resource sharing.

        University of California Los Angeles (UCLA) is awarded contract to create the Network Measurement Center.

        Bolt Beranek and Newman, Inc. (BBN) is awarded contract to build the Interface Message Processors (IMPs).

        Network Working Group (NWG), headed by Steve Crocker, is organized to develop host-level protocols for the ARPANET.

1969     The ARPANET is commissioned by the Department of Defense for research into networking.

        The first IMP (node 1) is installed at UCLA in September. It will function as the Network Measurement Center.

        The second IMP (node 2) is installed at Stanford Research Institute (SRI) in October. It will function as the Network Information Center.

        The third IMP (node 3) is installed at University of California Santa Barbara (UCSB) in November.

        The fourth IMP (node 4) is installed at University of Utah in December.

        The first packets are sent across the network, resulting in the first network crash.

        First Request For Comments (RFC) document, "Host Software," is written and distributed by Steve Crocker.

1970     ARPANET hosts begin using the first host-to-host protocol, Network Control Protocol (NCP).

        ALOHANET, the first packet radio network, is developed at the University of Hawaii by Norman Abramson.

        AT&T installs the first cross-country 56 Kbps links.

        The experimental packet-switched network at NPL in the United Kingdom becomes fully operational.

1971    ARPANET reaches initial target size of 15 nodes.

BBN develops Terminal IMP (TIP) to support up to 64 directly connected terminals into the ARPANET.

First networked email program is created by Ray Tomlinson of BBN.

Michael Hart starts Project Gutenberg to make copyright-free works available electronically. The first text is the U.S. Declaration of Independence.

1972    ARPA is renamed DARPA.

First computer-to-computer chat occurs at UCLA.

International Conference on Computer Communications provides first public demonstration of ARPANET, organized by Robert Kahn.

Ray Tomlinson of BBN sends first network email message across the ARPANET using SNDMSG.

France constructs its own packet-switched network called Cyclades.

RFC 318 defines TELNET.

1973    First international connections to the ARPANET, between University College of London (United Kingdom) and the Royal Radar Establishment (Norway).

Bob Kahn and Vinton Cerf begin engineering the architecture for the ARPA Internet Project.

Network Voice Protocol (NVP) enables conference calls over the ARPANET.

SRI starts publishing ARPANET News; ARPANET users estimated at 2,000.

Study shows 75% of ARPANET traffic consists of email.

RFC 454 defines File Transfer Protocol (FTP).

1974    The first internetworking protocol, TCP, is presented in a paper published by Kahn and Cerf, "A Protocol for Packet Network Intercommunication."

The first public, commercial packet-switched network, TELENET, starts operation.

1975    Defense Communications Agency (DCA) takes over ARPANET operation.

Satellite links now span the Atlantic and Pacific as first TCP tests are run.

EIES conferencing system goes online.

Steve Walker creates first ARPANET mailing list, MsgGroup.

1976    Queen Elizabeth II of the United Kingdom sends out her first email message.

Computer conferencing and interaction is explored: Jacques Vallee, Arthur C. Hastings, and Gerold Askevold, "Remote Viewing Experiments through Computer Conferencing;" Jacques Vallee, "The FORUM Project: Network Conferencing and its Future Applications;" Roxanne Hiltz and Murray Turoff, "Meeting Through Your Computer."

1977    First demonstration of interconnected networks, or Internet.

THEORYNET is developed at the University of Wisconsin by Larry Landweber to provide electronic mail to over 100 researchers via TELENET.

RFC 733 defines Mail protocol.

1978    TCP is divided into TCP and IP, collectively known as TCP/IP.

J. C. R. Licklider and Albert Vezza publish, "Applications of Information Networks."

Lawrence G. Roberts publishes, "The Evolution of Packet Switching."

1979    ARPA establishes the Internet Configuration Control Board (ICCB).

        Kevin MacKenzie emails the MsgGroup suggesting the use of emoticons in email to better convey an email's sentiments. The idea is not well received.

1980    An unintentional mishap in the form of a status message virus causes the ARPANET to effectively shut down on October 27.

1981    The Computer Science Network, CSNET, is built, funded through NSF.

        RFC 788 defines the Simple Mail Transfer Protocol (SMTP).

        Jacques Vallee publishes, "The User Abusers — Modern Networks in Perspective."

1982    The TCP/IP protocol suite is officially adopted by ARPA and specified to replace NCP.

1983    January 1 marks the cutover from NCP to TCP on the ARPANET.

        CSNET/ARPANET gateway is established.

        DCA splits off MILNET from the ARPANET with 45 nodes, leaving 68 nodes on the ARPANET.

1984    NSF takes over backbone administration from ARPA.

        Domain Name System (DNS) is developed.

        Number of (what is now Internet) hosts exceeds 1,000.

        Neuromancer written by William Gibson. In this book, Gibson coins the phrase "cyberspace."

        Brock N. Meeks, "An Overview of Conferencing Systems."

1985    DCA contracts responsibility for DNS root management to the Information Sciences Institute (ISI) at USC and for DNS NIC registrations to SRI.

        Symbolics.com is assigned the first registered domain. Other firsts include: cmu.edu, purdue.edu, rice.edu, berkeley.edu, ucla.edu, rutgers.edu, bbn.com, think.com, css.gov, mitre.org, .uk.

1986    IPTO ceases to exist as a discrete office after DARPA reorganization. Technical scope of IPTO expands and it becomes the Information Science and Technology Office (ISTO).

NSF develops NSFNET backbone, connecting five supercomputer centers and interconnecting other Internet sites at 56 Kbps.

Number of Internet hosts exceeds 10,000.

1987    Merit Network, Inc., IBM, and MCI sign agreement to manage the NSFNET backbone.

1988    NSFNET backbone upgraded to T1 (1.54 Mbps).

Internet Relay Chat (IRC) developed by Jarkko Oikarinen.

1989    First commercial electronic mail carriers appear on the Internet: MCI Mail and Compuserve.

Clifford Stoll publishes, "A Cuckoo's Egg," a first-hand account of an investigation into a German cracker group that infiltrated the ARPANET and other networked host computers.

Number of Internet hosts exceeds 100,000.

1990    ARPANET is officially decommissioned.

# Internet Milestones

The following table on Internet growth picks up where the corresponding table in the section above leaves off. Note how the exponential growth in the 1990s, shortly after the Internet starts to allow commercial traffic and the World Wide Web is introduced, coincides with a dramatic increase in the milestones for those years.

| Internet Growth | | | |
|---|---|---|---|
| Date | Hosts | Networks | Domains |
| 07/89 | 130,000 | 650 | 3,900 |
| 10/89 | 159,000 | 837 | |
| 10/90 | 313,000 | 2,063 | 9,300 |
| 01/91 | 376,000 | 2,338 | |
| 07/91 | 535,000 | 3,086 | 16,000 |
| 10/91 | 617,000 | 3,556 | 18,000 |
| 01/92 | 727,000 | 4,526 | |
| 04/92 | 890,000 | 5,291 | 20,000 |
| 07/92 | 992,000 | 6,569 | 16,300 |
| 10/92 | 1,136,000 | 7,505 | 18,100 |
| 01/93 | 1,313,000 | 8,258 | 21,000 |
| 04/93 | 1,486,000 | 9,722 | 22,000 |
| 07/93 | 1,766,000 | 13,767 | 26,000 |
| 10/93 | 2,056,000 | 16,533 | 28,000 |
| 01/94 | 2,217,000 | 20,539 | 30,000 |
| 07/94 | 3,212,000 | 25,210 | 46,000 |
| 10/94 | 3,864,000 | 37,022 | 56,000 |
| 01/95 | 4,852,000 | 39,410 | 71,000 |
| 07/95 | 6,642,000 | 61,538 | 120,000 |
| 01/96 | 9,472,000 | 93,671 | 240,000 |
| 07/96 | 12,881,000 | 134,365 | 488,000 |
| 01/97 | 16,146,000 | 828,000 | |
| 07/97 | 19,540,000 | 1,301,000 | |

1972    The InterNetworking Group (INWG) is formed to establish networking protocols.

BBN forms Telenet Communications Corporation to develop private packet-switched network systems.

RFC 318 defines TELNET.

1973    Bob Kahn and Vinton Cerf begin engineering the architecture for the ARPA Internet Project.

Bob Metcalfe presents his initial description of Ethernet in his Harvard Ph.D. thesis and creates the first Ethernet-based network at Xerox PARC.

RFC 454 defines FTP.

1974    The first internetworking protocol, TCP, is presented in a paper published by Kahn and Cerf, "A Protocol for Packet Network Intercommunication."

The first public, commercial packet-switched network, TELENET, starts operation.

Scrapbook information processing system developed for network at NPL by a team led by David Yates. It includes a database, word processing, and network email.

1975    Operational control of the ARPANET, along with the Internet project, is transferred to the Defense Communications Agency (DCA).

1976    AT&T Bell Laboratories develops UUCP (Unix-to-Unix Copy); it starts distributing it freely in 1977.

1977    The first Internet demonstration occurs, proving the viability of interconnecting different network types through the Internet protocols and specialized gateway routers.

Additional private networks are launched, including THEORYNET and Tymnet.

RFC 733 defines Mail protocol.

1978    TCP is split into TCP and IP.

RCPM dial-in Bulletin Board System (BBS) started by Ward Christianson and Randy Suess.

1979    ARPA creates Internet Configuration Control Board (ICCB).

USENET begins operating between Duke and University of North Carolina.

First Multi-User Dungeon (MUD) interactive gaming, MUD1, is started up in England.

1981    The BITNET cooperative network begins operating.

The Computer Science Network (CSNET) is built, funded through NSF.

RFC 788 defines the Simple Mail Transfer Protocol (SMTP).

1982    The TCP/IP protocol suite is officially adopted by ARPA and DCA for use on the ARPANET, and becomes part of the definition for the Internet.

The U.S. Department of Defense specifies TCP/IP as their standard networking protocols.

France Telecom introduces its Minitel system, becoming the first phone company to offer a network service into people's homes combining content with telecommunications.

European Unix Network (EUNET) is established to provide email and USENET services.

RFC 827 defines the External Gateway Protocol (EGP)

1983    January 1 marks the cutover from NCP to TCP on the ARPANET.

CSNET/ARPANET gateway is established.

DCA splits MILNET off of ARPANET with 45 nodes, leaving 68 nodes on the ARPANET.

Desktop workstations are introduced; many are network-ready. Berkeley releases 4.2BSD, incorporating TCP/IP.

Internet Activities Board (IAB) is established, replacing the ICCB.

European Academic and Research Network (EARN) is established; similar to BITNET, with a gateway funded by IBM.

FidoNet BBS software developed by Tom Jennings. Supports the growth of a hobbyist electronic mail and conferencing network

1984    Domain Name System (DNS), developed by Jon Postel, Paul Mockapetris, and Craig Partridge, is brought into operation.

Number of Internet hosts exceeds 1,000.

Japan Unix Network (JUNET) established using UUCP.

Joint Academic Network (JANET) is established in the U.K. using the Coloured Book protocols; previously SERCNET.

Moderated newsgroups are introduced on USENET (mod.*).

1985    Whole Earth 'Lectronic Link (WELL) started.

Information Sciences Institute (ISI) at USC is given responsibility for DNS root management by DCA, and SRI for DNS NIC registrations.

Symbolics.com is assigned on 15 March to become the first registered domain. Other firsts: cmu.edu, purdue.edu, rice.edu, ucla.edu (April); css.gov (June); mitre.org, .uk (July).

1986    NSF develops NSFNET backbone, connecting five supercomputer centers and interconnecting other Internet sites at 56 Kbps.

Internet Engineering Task Force (IETF) and Internet Research Task Force (IRTF) are established under the IAB.

The first Freenet (Cleveland) becomes operational under the auspices of the Society for Public Access Computing (SoPAC).

The Network News Transfer Protocol (NNTP) is developed to enhance USENET performance over TCP/IP.

Mail Exchanger (MX) records are developed by Craig Partridge to allow non-IP network hosts to have domain addresses.

Number of Internet hosts exceeds 10,000.

1987    Merit Network, Inc., IBM, and MCI sign agreement to manage the NSFNET backbone.

Bay Area Regional Research Network (BARRNET) becomes operational.

UUNET is founded with Usenix funds to provide

commercial UUCP and USENET access. Originally an experiment by Rick Adams and Mike O'Dell.

Email link is established between Germany and China using the CSNET protocols, with the first message from China sent on 20 September.

RFC 1000, "Request for Comments Reference Guide," is published.

1988    NSFNET backbone upgraded to T1 (1.54 Mbps).

U.S. Department of Defense adopts OSI and considers use of TCP/IP as an interim solution. U.S. Government OSI Profile (GOSIP) defines the set of protocols to be supported by Government purchased products.

Los Nettos network created with no federal funding; supported by regional members (founding: Caltech, TIS, UCLA, USC, ISI).

California Education and Research Federation network (CERFNET) founded by Susan Estrada.

First Canadian regionals join NSFNET: ONet via Cornell, RISQ via Princeton, BCnet via University of Washington.

Internet Relay Chat (IRC) developed by Jarkko Oikarinen.

The Morris Internet worm is unleashed.

Computer Emergency Response Team (CERT) is formed by DARPA in response to the needs exhibited during the Morris worm incident. The worm is the only advisory issued this year.

First gateway connecting FidoNet and the Internet, enabling email and news to be exchanged.

Countries connecting to the NSFNET: Canada (CA), Denmark (DK), Finland (FI), France (FR), Iceland (IS), Norway (NO), Sweden (SE).

1989    First commercial electronic mail carriers appear on the Internet: MCI Mail and Compuserve.

Reseaux IP Europeens (RIPE) formed by European service providers to ensure the necessary administrative and technical coordination to allow the operation of the pan European IP Network.

Corporation for Research and Education Networking (CREN) is formed by merging CSNET into BITNET.

Number of Internet hosts exceeds 100,000.

7 CERT advisories are issued.

Countries connecting to the NSFNET: Australia (AU), Germany (DE), Israel (IL), Italy (IT), Japan (JP), Mexico (MX), Netherlands (NL), New Zealand (NZ), Puerto Rico (PR), United Kingdom (UK).

1990    ARPANET officially decommissioned.

Electronic Frontier Foundation (EFF) founded by Mitch Kapor.

Archie released by Peter Deutsch, Alan Emtage, and Bill Heelan at McGill.

Australian Academic Research Network (AARNET) becomes operational.

CA*NET formed by 10 regional networks as national Canadian backbone with direct connection to the NSFNET.

The first remotely operated machine to be hooked up to the Internet, the Internet Toaster, debuts at Interop.

12 CERT advisories and 130 reports are issued.

Countries connecting to NSFNET: Argentina (AR), Austria (AT), Belgium (BE), Brazil (BR), Chile (CL), Greece (GR), India (IN), Ireland (IE), Korea (KR), Spain (ES), Switzerland (CH).

1991    NSF removes restrictions on commercial use of the Internet.

High Performance Computing Act, authored by then Senator Gore, is signed into law.

Commercial Internet eXchange (CIX) Association, Inc. formed by General Atomics (CERFnet), Performance Systems International, Inc. (PSInet), and UUNET Technologies, Inc. (AlterNet), after NSF removes restrictions on commercial use of the Internet.

World Wide Web software created by Tim Berners-Lee is released by CERN, the European Laboratory for Particle Physics.

Wide Area Information Servers (WAIS), invented by Brewster Kahle, is released by Thinking Machines Corporation.

Gopher is released by Paul Lindner and Mark P. McCahill from the University of Minnesota.

Pretty Good Privacy (PGP) is released by Philip Zimmerman.

U.S. High Performance Computing Act (Gore 1) establishes the National Research and Education Network (NREN).

NSFNET backbone upgraded to T3 (44.736 Mbps), traffic exceeds 1 trillion bytes/month and 10 billion packets/month.

JANET IP Service (JIPS) starts operation, signaling the changeover from Coloured Book software to TCP/IP within the UK academic network. IP was initially tunneled within X.25.

CERT issues 23 advisories.

Countries connecting to the NSFNET: Croatia (HR), Czech Republic (CZ), Hong Kong (HK), Hungary (HU), Poland (PL), Portugal (PT), Singapore (SG), South Africa (ZA), Taiwan (TW), Tunisia (TN).

1992     The Internet Society (ISOC) is chartered.

The number of Internet hosts exceeds 1,000,000.

The first MBONE audio multicast (March) and video multicast (November) take place.

The RIPE Network Coordination Center (NCC) is created to provide address registration and coordination services to the European Internet community.

The IAB is reconstituted as the Internet Architecture Board and becomes part of the Internet Society.

Veronica, a gopherspace search tool, is released by University of Nevada.

The World Bank comes on-line.

Japan's first ISP, Internet Initiative Japan (IIJ), is formed by Koichi Suzuki.

The term "Surfing the Internet" is coined by Jean Armour Polly.

Internet Hunt started by Rick Gates.

CERT issues 21 advisories and 800 reports.

Countries connecting to the NSFNET: Antarctica (AQ), Cameroon (CM), Cyprus (CY), Ecuador (EC), Estonia (EE), Kuwait (KW), Latvia (LV), Luxembourg (LU), Malaysia (MY), Slovakia (SK), Slovenia (SI), Thailand (TH), Venezuela (VE).

1993     The InterNIC is created by the NSF to provide specific Internet services, including: directory and database services (AT&T), registration services (Network Solutions Inc.), information services (General Atomics/CERFnet).

The U.S. White House comes on-line with a Web site and email addresses for President Bill Clinton and Vice President Al Gore.

New types of worms invade the Internet: Web Worms (W4), Spiders, Wanderers, Crawlers, and Snakes, etc.

Internet Talk Radio begins broadcasting.

The United Nations comes on-line.

U.S. National Information Infrastructure Act.

Mosaic takes the Internet by storm; the Web proliferates at a 341,634% annual growth rate of service traffic. Gopher's growth is 997%.

CERT issues 18 advisories and 1300 reports.

Countries connecting to NSFNET: Bulgaria (BG), Costa Rica (CR), Egypt (EG), Fiji (FJ), Ghana (GH), Guam (GU), Indonesia (ID), Kazakhstan (KZ), Kenya (KE), Liechtenstein (LI), Peru (PE), Romania (RO), Russian Federation (RU), Turkey (TR), Ukraine (UA), UAE (AE), US Virgin Islands (VI).

1994    The U.S. Senate and House provide online information servers.

Shopping malls arrive on the Internet; Pizza Hut provides online ordering; and the first cyberbank, First Virtual, starts operating on the Internet.

The first cyberstation, RT-FM, broadcasts from Interop in Las Vegas; and existing radio stations start rebroadcasting their signals on the Internet.

The National Institute for Standards and Technology (NIST) suggests that GOSIP should incorporate TCP/IP and drop the OSI-only requirement.

Arizona law firm of Canter and Siegel spams the Internet with email advertising green card lottery services; Internet citizens flame back.

NSFNET traffic passes 10 trillion bytes/month.

The World Wide Web edges out TELNET to become the second most popular service on the Internet, based on the percentage of packets and bytes traffic on the NSFNET. FTP remains number one.

The Trans-European Research and Education Network

Association (TERENA) is formed by the merger of RARE and EARN, with representatives from 38 countries, as well as CERN and ECMWF. Their goal is to: "promote and participate in the development of a high quality international information and telecommunications infrastructure for the benefit of research and education."

CERT issues 15 advisories and 2300 reports.

Countries connecting to the NSFNET: Algeria (DZ), Armenia (AM), Bermuda (BM), Burkina Faso (BF), China (CN), Colombia (CO), Jamaica (JM), Lebanon (LB), Lithuania (LT), Macau (MO), Morocco (MA), New Caledonia, Nicaragua (NI), Niger (NE), Panama (PA), Philippines (PH), Senegal (SN), Sri Lanka (LK), Swaziland (SZ), Uruguay (UY), Uzbekistan (UZ).

1995   NSFNET reverts back to a research network. The main U.S. backbone traffic is now routed through commercial, interconnected network providers.

The new NSFNET is born as NSF establishes the very-high-speed Backbone Network Service (vBNS) linking super computing centers.

Hong Kong police disconnect all but 1 of the colony's Internet providers in search of a hacker. 10,000 people are left without Internet access.

RealAudio, an audio streaming technology, brings real-time audio to the Internet.

Radio HK, the first 24 hour Internet only radio station starts broadcasting.

In March, The World Wide Web surpasses FTP data as the service with the greatest traffic.

Traditional online dial-up systems (Compuserve, America Online, Prodigy) begin to provide Internet access.

A number of Internet-related companies go public, with Netscape leading the pack with the third largest ever NASDAQ IPO share value.

Thousands in Minneapolis-St. Paul area in the U.S. lose Internet access after transients start a bonfire under a bridge at the University of Minnesota causing fiber optic cables to melt.

Registration of domain names is no longer free. Beginning 14 September, a $50 annual fee has been imposed, which up until then was subsidized by NSF. NSF continues to pay for .edu registration and, on an interim basis, for .gov.

The Vatican comes online.

The Canadian government comes online.

The first official Internet wiretap was successful in helping the Secret Service and Drug Enforcement Agency (DEA) apprehend three individuals who were illegally manufacturing and selling cell phone cloning equipment and electronic devices.

Operation Home Front connects, for the first time, soldiers in the field with their families back home via the Internet.

Richard White becomes the first person to be declared a munition, under the USA's arms export control laws, because of an RSA file security encryption program emblazoned on his arm.

CERT issues 18 advisories and 2412 reports.

Country domains registered: Ethiopia (ET), Cote d'Ivoire (CI), Cook Islands (CK) Cayman Islands (KY), Anguilla (AI), Gibraltar (GI), Vatican (VA), Kiribati (KI), Kyrgyzstan (KG), Madagascar (MG), Mauritius (MU), Micronesia (FM), Monaco (MC), Mongolia (MN), Nepal (NP), Nigeria (NG), Western Samoa (WS), San Marino (SM), Tanzania (TZ), Tonga (TO), Uganda (UG), Vanuatu (VU).

1996    Number of Internet hosts exceeds 12.8 million. President Clinton and Vice-President Gore announce "Next Generation Internet" initiative.

1997    Internet phones catch the attention of U.S. telecommunication companies who ask the U.S. Congress to ban the technology (which has been around for years).

The controversial U.S. Communications Decency Act (CDA) becomes law in the U.S. in order to prohibit distribution of indecent materials over the Net. A few months later a three-judge panel imposes an injunction against its enforcement. Supreme Court unanimously rules most of it unconstitutional in 1997.

9,272 organizations find themselves unlisted after the InterNIC drops their name service as a result of not having paid their domain name fee.

Various ISPs suffer extended service outages, bringing into question whether they will be able to handle the growing number of users. AOL (19 hours), Netcom (13 hours), AT&T WorldNet (28 hours — email only).

New Yorks' Public Access Networks Corp (PANIX) is shut down after repeated SYN attacks by a cracker using methods outlined in a hacker magazine.

Various U.S. Government sites are hacked into and their content changed, including the CIA, Department of Justice, Air Force.

MCI upgrades Internet backbone adding ˜13,000 ports, bringing the effective speed from 155 Mbps to 622 Mbps.

The Internet Ad Hoc Committee announces plans to add 7 new generic Top Level Domains (gTLD):

A malicious cancelbot is released on USENET wiping out more than 25,000 messages.

The Web browser war, fought primarily between Netscape and Microsoft, has rushed in a new age in software development, whereby new releases are made quarterly with the help of Internet users eager to test upcoming (beta) versions.

Restrictions on Internet use around the world: China requires users and ISPs to register with the police;

Germany cuts off access to some newsgroups carried on Compuserve; Saudi Arabia confines Internet access to universities and hospitals; Singapore requires political and religious content providers to register with the State; New Zealand classifies computer disks as "publications" that can be censored and seized.

CERT issues 27 advisories and 2573 reports.

Country domains registered: Qatar (QA), Vientiane (LA), Djibouti (DJ), Niger (NE), Central African Republic (CF), Mauretania (MF), Oman (OM), Norfolk Island (NF), Tuvalu (TV), French Polynesia (PF), Syria (SY), Aruba (AW), Cambodia (KH), French Guiana (GF), Eritrea (ER), Cape Verde (CV), Burundi (BI), Benin (BJ) Bosnia-Hercegovina (BA), Andorra (AD), Guadeloupe (GP), Guernsey (GG), Isle of Man (IM), Jersey (JE), Lao (LA), Maldives (MV), Marshall Islands (MH), Mauritania (MR), Northern Mariana Islands (MP), Rwanda (RW), Togo (TG), Yemen (YE), Zaire (ZR).

# Netiquette

Communicating over the Internet, whether by email, chat, or some other means, is not likely to be well regarded by people who have high standards when it comes to spelling and grammar, or by those who fondly recall learning in school all about salutations and the various and sundry parts that compose a well formed paragraph. You can blame the keyboard or bad typing skills, or both. Or you can blame the innate sense of informality in the medium. Or you can attribute it to the immediacy of the environment and that feeling of urgency to send out a quick email or to keep pace in a fast moving, chat room conversation.

Netiquette, short for network etiquette, is a loose collection of rules and conventions applied to online behavior, particularly to communicating over the Internet. Most of what is commonly understood as netiquette — specific perspectives and formal descriptions vary considerably — represents common sense behavioral considerations, things we try to apply in our general interactions with others but adapted to the new and unique communication environment of the Internet. In general, these conventions focus on respecting other people's time and privacy, sharing knowledge rather than withholding it or lauding it above others, not abusing one's power (e.g., when acting as moderator in a chat room or newsgroup), and forgiving mistakes, especially with respect to new Internet denizens, also known as newbies. Netiquette also tries to cover some operational basics, like knowing when and how to ask for help, understanding that all uppercase letters make you appear as if you are shouting, thinking twice before responding emotionally (also known as flaming someone), not forwarding unsolicited mail to your friends, and so on.

The simplest and most fundamental netiquette rule is not letting the remoteness imposed by the technology allow you to forget that you are interacting with other people. Typing in your thoughts at a computer keyboard, in the isolation of your room or in the public space of an office or library, makes it easy to forget that someone else will eventually be reading those words; and they won't necessarily know what you were feeling when you wrote them, or precisely what you intended. You can't accompany your words with your facial expressions (well, see below, you can try). You can't easily or exactly communicate your tone of voice. You can't clearly identify when you're being serious, and when sarcastic. So, don't be surprised when you are misunderstood, and try to use whatever conventions you can, including those in the tables below, to assist you in communicating both your thoughts *and* the sentiments behind them. Equally important, apply the same conventions when reading what others have written. Do not presume you know exactly what your friend, business associate, or even your favorite in-law meant to convey when reading his or her words in an email message or in a chat window. Give them the benefit of the doubt, because the technology only goes so far.

If you want to communicate over the Internet using some of the language constructs that the Internet has inspired and propagated, or if you just want to better understand the common acronyms and expressions that commonly punctuate much of the Internet's communication traffic, you'll find the following tables of assistance.

| Internet Shorthand Acronyms | |
|---|---|
| Acronym | Meaning |
| AFAIK | As Far As I Know |
| AFK | Away From Keyboard |
| AOLer | America OnLine Member |
| A/S/L | Age/Sex/Location |
| BAK | Back At Keyboard |
| BBIAF | Be Back In A Flash |
| BBL | Be Back Later |
| BD | Big Deal |
| BFD | Big Friggin' Deal |
| BFN | Bye For Now |
| BRB | Be Right Back |
| BTW | By The Way |
| CUL8R | See You Later |
| CYA | See Ya |
| FB | Furrowed Brow |
| FWIW | For What It's Worth |
| GDM8 | G'day Mate |
| GMTA | Great Minds Think Alike |
| GRD | Grinning, Running, Ducking |
| GR8 | Great |
| HTH | Hope This Helps |
| IAE | In Any Event |
| IANAL | I Am Not A Lawyer |
| IM | Instant Message |
| IMHO | In My Humble Opinion |
| IMNSHO | In My Not So Humble Opinion |
| IOW | In Other Words |
| IYSWIM | If You See What I Mean |
| J/K | Just Kidding |
| LMAO | Laughing My A-- Off |
| LOL | Laughing Out Loud |
| LTNS | Long Time No See |
| M4M | Men seeking Men |
| NFW | No Friggin' Way |
| NP | No Problem |
| NRN | No Reply Necessary |

| Internet Shorthand Acronyms | |
|---|---|
| Acronym | Meaning |
| NW | No Way |
| OIC | Oh, I See |
| OTOH | On The Other Hand |
| PBT | Pay Back Time |
| ROTFL/ROFL | Rolling On The Floor Laughing |
| RTFM | Read The Friggin' Manual |
| SOL | Sooner Or Later |
| TOS | Terms Of Service |
| TTFN | Ta-Ta For Now |
| TTYL | Talk To You Soon |
| WB | Welcome Back |
| WTG | Way To Go |
| YL/YM | Young Lady/Young Man |
| YMMV | Your Mileage May Vary |

| Internet Shorthand Expressions | |
|---|---|
| Expression | Meaning |
| O:-) | Angel |
| *^_^* | Big Grin |
| T_T | Big Tears |
| @^_^@ | Blushing |
| :'-( | Crying |
| }:>; | Devil |
| :-e | Disappointed |
| :-L~~ | Drooling |
| X= | Fingers Crossed |
| :-( | Frowning |
| $-) | Greedy |
| 8:)3)= | Happy Girl |
| {{{{Whomever}}}} | Hug for Whomever |
| {} | Hugs |
| X-) | I See Nothing |
| :-X | I'll Say Nothing |
| ****** | Kisses |
| :* | Kissing |

| Internet Shorthand Expressions | |
|---|---|
| Expression | Meaning |
| : = ) | Little Hitler |
| : - D | Laughing |
| @ ] ' - , - - - - - | Rose |
| : - @ | Screaming |
| : - O | Shock |
| : - ) | Smiling |
| : - P | Sticking Out Tongue |
| ^_^; | Sweating |
| (hmm) Ooo . . : - ) | Thinking Happy Thoughts |
| (hmm) Ooo . . : - ( | Thinking Sad Thoughts |
| ; - ) | Winking |
| \ \ / / | Vulcan Salute |

# Common Internet Age Jargon

The following list contains a sampling of the jargon inspired by the Web, the Internet, and the ever growing pervasiveness of computers in our lives.

404      Someone who's clueless. From the Web error message "404, URL Not Found," meaning that the document you've tried to access can't be located. "Don't bother asking him...he's 404, man."

Adminisphere   The rarefied organizational layers beginning just above the rank and file. Decisions that fall from the adminisphere are often profoundly inappropriate or irrelevant to the problems they were designed to solve.

Alpha Geek    The most knowledgeable, technically proficient person in an office or work group. "Ask Larry, he's the alpha geek around here."

| | |
|---|---|
| Assmosis | The process by which some people seem to absorb success and advancement by kissing up to the boss rather than working hard. |
| Beepilepsy | The brief seizure people sometimes suffer when their beepers go off, especially in vibrator mode. Characterized by physical spasms, goofy facial expressions, and stopping speech in mid-sentence. |
| Blamestorming | Sitting around in a group discussing why a deadline was missed or a project failed, and who was responsible. |
| Bookmark | To take note of a person for future reference (a metaphor borrowed from web browsers). "I bookmarked him after seeing his cool demo at Siggraph." |
| Blowing Your Buffer | Losing one's train of thought. Occurs when the person you are speaking with won't let you get a word in edgewise or has just said something so astonishing that your train gets derailed. "Damn, I just blew my buffer!" |
| Career-Limiting Move (CLM) | Used among microserfs to describe an ill-advised activity. Trashing your boss while he or she is within earshot is a serious CLM. |
| CGI Joe | A hard-core CGI script programmer with all the social skills and charisma of a plastic action figure. |
| Chainsaw Consultant | An outside expert brought in to reduce the employee headcount, leaving the top brass with clean hands. |

| | |
|---|---|
| Chips and Salsa | Chips hardware, salsa software. "Well, first we gotta figure out if the problem's in your chips or your salsa." |
| Chip Jewelry | A euphemism for old computers destined to be scrapped or turned into decorative ornaments. "I paid three grand for that Mac SE, and now it's nothing but chip jewelry." |
| Circling The Drain | Used to describe projects that have no more life in them but refuse to die. "That disk conversion project has been circling the drain for years." |
| Cobweb Site | A Web site that hasn't been updated for a long time. A dead web page. |
| Crapplet | A badly written or profoundly useless Java applet. "I just wasted 30 minutes downloading this stinkin' crapplet!" |
| Crash Test Dummies | Those of us who pay for unstable, not-yet-ready-for-prime-time software foisted on us by computer companies. |
| Critical Mess | An unstable stage in a software project's life in which any single change or bug fix can result in the creation of two or more new bugs. Continued development at this stage can lead to an exponential increase in the number of bugs. |
| Cube Farm | An office filled with cubicles. |
| Dancing Baloney | Little animated GIFs and other Web F/X that are useless and serve simply to impress clients. "This page is kinda dull. Maybe a little dancing baloney will help." |
| Dawn Patrol | Programmers who are still at their terminals when the day shift returns to work the next morning. Usually found in Trog Mode (see below). |

| | |
|---|---|
| Dead Tree Edition | The paper version of a publication available in both paper and electronic forms, as in: "The dead tree edition of the San Francisco Chronicle..." |
| Depotphobia | Fear associated with entering a Home Depot because of how much money one might spend. Electronics geeks experience Shackophobia. |
| Dilberted | To be exploited and oppressed by your boss. Derived from the experiences of Dilbert, the geek-in-hell comic strip character. "I've been dilberted again. The old man revised the specs for the fourth time this week." |
| Domain Dipping | Typing in random words between www. and .com just to see what's out there. |
| Dorito Syndrome | Feelings of emptiness and dissatisfaction triggered by addictive substances that lack nutritional content. "I just spent six hours surfing the Web, and now I've got a bad case of Dorito Syndrome." |
| Dustbuster | A phone call or email message sent to someone after a long while just to "shake the dust off" and see if the connection still works. |
| Egosurfing | Scanning the net, databases, print media, or research papers looking for the mention of your name. |
| Elvis Year | The peak year of something's popularity. "Barney the dinosaur's Elvis year was 1993." |
| Email Tennis | When you email someone who responds while you are still answering mail. You respond again, and so forth, as if you were |

carrying on a chat via email messages. "Ok, enough of this email tennis, why don't I call you?"

| | |
|---|---|
| Flight Risk | Planning to leave a company or department soon. |
| Generica | Features of the American landscape that are exactly the same no matter where one is. "We were so lost in generica, I actually forgot what city we were in." |
| Future-Proof | Term used to describe a technology that supposedly won't become technologically outdated (at least anytime soon). |
| Glazing | Corporate-speak for sleeping with your eyes open. A popular pastime at conferences and early-morning meetings. "Didn't he notice that half the room was glazing by the second session?" |
| Going Cyrillic | When a graphical display (LED or LCD screen, monitor, etc.) starts to display garbage. "The thing just went cyrillic on me." |
| GOOD Job | A "Get-Out-Of-Debt" job. A well paying job people take in order to pay off their debts, one that they will quit as soon as they are solvent again. |
| Gray Matter | Older, experienced business people hired by young entrepreneurial firms looking to appear more reputable and established. |
| Graybar Land | The place you go while you're staring at a computer that's processing something very slowly (while you watch the gray bar creep across the screen). "I was in graybar land for what seemed like hours, thanks to that CAD rendering." |

| | |
|---|---|
| Hourglass Mode | Waiting in limbo for some expected action to take place. "I was held up at the post office because the clerk was in hourglass mode." |
| Idea Hamsters | People who always seem to have their idea generators running. "That guy's a real idea hamster. Give him a concept and he'll turn it over 'til he comes up with something useful." |
| IQueue | The line of interesting email messages waiting to be read after one has deleted all of the junk mail. |
| Irritainment | Entertainment and media spectacles that are annoying, but you find yourself unable to stop watching them. The O.J. trials were a prime example. |
| It's a Feature | From the adage "It's not a bug, it's a feature." Used sarcastically to describe an unpleasant experience that you wish to gloss over. |
| Keyboard Plaque | The disgusting buildup of dirt and crud found on computer keyboards. "Are there any other terminals I can use? This one has a bad case of keyboard plaque." |
| Link Rot | The process by which links on a web page became as obsolete as the sites they're connected to change location or die. |
| Martian Mail | An email that arrives months after it was sent (as if it has been routed via Mars). |
| Meatspace | The physical world (as opposed to the virtual) also carbon community, facetime, F2F, RL. |
| Midair Passenger Exchange | Grim air-traffic controller-speak for a head-on collision. Midair passenger |

exchanges are quickly followed by "aluminum rain."

Monkey Bath

A bath so hot that, when lowering yourself in, you go "Oo! Oo! Oo! Ah! Ah! Ah!."

Mouse Potato

The online, wired generation's answer to the couch potato.

Notwork

A network in its non-working state.

Nyetscape

Nickname for AOL's less-than-full-featured Web browser.

Ohnosecond

That miniscule fraction of time in which you realize that you've just made a BIG mistake. Seen in Elizabeth P. Crowe's book, "The Electronic Traveller."

Open-Collar Workers

People who work at home or telecommute.

PEBCAK

Tech support shorthand for "Problem Exists Between Chair and Keyboard." Another variation on the above is ID10T: "This guy has an ID-Ten-T on his system."

Percussive Maintenance

The fine art of whacking the crap out of an electronic device to get it to work again.

Plug-and-Play

A new hire who doesn't need any training. "The new guy, John, is great. He's totally plug-and-play."

Prairie Dogging

When someone yells or drops something loudly in a "cube farm" (an office full of cubicles) and everyone's heads pop up over the walls to see what's going on.

Print Mile

The distance covered between a desk and a printer shared by a group of users in an office. "I think I've traveled enough print miles on this job to qualify for a vacation."

| | |
|---|---|
| Salmon Day | The experience of spending an entire day swimming upstream only to get screwed in the end. |
| Seagull Manager | A manager who flies in, makes a lot of noise, craps over everything and then leaves. |
| Shovelware | A Web document that was shoveled from paper onto the Web, help system, or whatever without much effort to adapt it to the new medium. Betrayed by, among other things, papercentric phrases like "See page so-and-so," "later in this booklet," and so forth. |
| SITCOMs | What yuppies turn into when they have children and one of them stops working to stay home with the kids. Stands for Single Income, Two Children, Oppressive Mortgage. |
| Square-headed Girlfriend | Another word for a computer. The victim of a square-headed girlfriend is a "computer widow." |
| Squirt The Bird | To transmit a signal up to a satellite. "Crew and talent are ready...what time do we squirt the bird?" |
| Starter Marriage | A short-lived first marriage that ends in a divorce with no kids, no property and no regrets. |
| Stress Puppy | A person who seems to thrive on being stressed out and whiny. |
| Swiped Out | An ATM or credit card that has been rendered useless because the magnetic strip is worn away from extensive use. |

Telephone Number Salary
A salary (or project budget) that has seven digits.

Thrashing
Clicking helter-skelter around an interactive computer screen or Web site in search of hidden buttons or links that might trigger actions.

Tourists
People who are taking training classes just to get a vacation from their jobs. "We had about three serious students in the class; the rest were tourists."

Treeware
Hacker slang for documentation or other printed material.

Triple-dub
An abbreviated way of saying www when speaking about a URL. "Check out this cool web site at triple-dub dot enlightenment dot co dot uk."

Trog Mode
A round-the-clock computer session in which your eyes get so tired you have to turn off the lights and toggle the monitor into reverse — white letters on a black screen. Often used at Dawn Patrol period (see above).

Umfriend
A sexual relation of dubious standing. "This is Dale, my...um...friend..."

Under Mouse Arrest
Getting busted for violating an online service's rule of conduct. "Sorry I couldn't get back to you. AOL put me under mouse arrest."

Uninstalled
Euphemism for being fired. Heard on the voicemail of a vice president at a downsizing computer firm: "You have reached the number of an uninstalled vice president. Please dial our main number and ask the operator for assistance." Also known as Decruitment.

| | |
|---|---|
| Voice Jail System | A poorly designed voicemail system that has so many submenus that one gets lost and has to hang up and call back. |
| Vulcan Nerve Pinch | The taxing hand position required to reach all of the appropriate keys for certain commands. For instance, the warm boot for a Mac II involves simultaneously pressing the Control key, the Command key, the Return key and the Power On key. |
| World Wide Wait | The real meaning of WWW. |
| Yuppie Food Stamps | The ubiquitous $20 bills spewed out of ATMs everywhere. Often used when trying to split the bill after a meal: "We all owe $8 each, but all anybody's got is yuppie food stamps." |

# Notes

## 1. Computers: The Instrument of a Revolution

1   Geoffrey D. Austrian, "Herman Hollerith: Forgotten Giant of Information Processing," Columbia University Press, 1982, 16.

2   Ibid, 9.

3   Ibid, 69.

4   F. H. Hinsley, Alan Stripp, "Code Breakers: The Inside Story of Bletchley Park," Oxford University Press, 1994, 134.

5   William Aspray, "John von Neumann and the Origins of Modern Computing," The MIT Press, 1990, 25.

6   Ibid, 34.

7   Ibid, 39.

8   T. R. Reid, "The Chip: How Two Americans Invented the Microchip and Launched a Revolution," Random House, 2001, 11.

9   Ibid, 22-23.

10  James Gillies, Robert Cailliau, "How the Web was Born: the Story of the World Wide Web," Oxford University Press, 2000, 116.

11    Reid, "The Chip," 177.


# 2. The ARPANET: A Network is Born

1    U.S. Department of Defense Web Site, "DARPA Over the Years," http://www.darpa.mil/body/overtheyears.html, 2002.

2    Gillies, Cailliau, "How the Web was Born," 14.

3    Ibid, 15.

4    Ibid, 16.

5    Janet Abbate, "Inventing the Internet," The MIT Press, 2000, 37-38.

6    Ibid, 46.

7    Ibid, 43.

8    Ibid, 56.

9    Ibid, 57.

10   Ibid, 60-61.

11   Ibid, 62-63.

12   Ibid, 49-50.

13   Gillies, Cailliau, "How the Web was Born," 28-29.

14   Abbate, "Inventing the Internet," 64.

15   Gillies, Cailliau, "How the Web was Born," 30.

16   Ibid, 29.

17   Ibid, 32.

18   Abbate, "Inventing the Internet," 79.

19   Gillies, Cailliau, "How the Web was Born," 33.

20   Ibid, 93.

21   Ibid, 115.

22   Gillies, Cailliau, "How the Web was Born," 33.

23   Abbate, "Inventing the Internet," 118.

24   Ibid, 120.

25   Ibid, 122.

26   Ibid, 122.

27   Ibid, 124-125.

28   Gillies, Cailliau, "How the Web was Born," 42.

29   Abbate, "Inventing the Internet," 128-129.

30   Ibid, 129.

31   Ibid, 130.

32   Gillies, Cailliau, "How the Web was Born," 80-81.

33   Abbate, "Inventing the Internet," 135-136.

34   Ibid, 138.

35   Ibid, 140-142.

36   Ibid, 50.

37   Ibid, 186.

38   The chart is courtesy of Dr. Lawrence Roberts. Any errors in representing the data are mine.

39   Abbate, "Inventing the Internet," 194.

40   Ibid, 195.

# 3. The Internet: A Network of Networks

1   Robert E. Kahn, Vinton G. Cerf, "What is the Internet (And What Makes it Work)," Briefing to the President, http://www.internetpolicy.org, 1999, 15.

2    Ibid, 2.

3    Gillies, Cailliau, "How the Web was Born," 70-71.

4    Abbate, "Inventing the Internet," 192.

5    Ibid, 193-194.

6    B. Kahin, "Commercialization of the Internet, Summary Report," RFC-1192, November 1990.

7    Ibid.

8    Office of Science and Technology Policy, "The Federal High Performance Computing Program," September 8, 1989, 35.

9    Kahin, "Commercialization of the Internet".

10   Abbate, "Inventing the Internet," 199.

11   Ibid, 198.

12   Ibid, 199.

13   Ibid, 199.

14   Ibid, 169.

15   Gillies, Cailliau, "How the Web was Born," 65.

# 4. Packet Switching: Lifeblood of the Internet

1    Paul Baran, "On Distributed Computing," Rand Corporation, 1964.

2    Abbate, "Inventing the Internet," 11.

3    Ibid, 124.

4    Ibid, 126.

# 5. Protocols: The Definition of Interoperability

1    Abbate, "Inventing the Internet," 130.

2    Ibid, 141.

General Sources

Defense Advanced Research Projects Agency, Information Processing Techniques Office, "Internet Protocol, RFC 791," September 1981.

R. Braden, "Requirements for Internet Hosts -- Communications Layers, RFC 1122," October 1989.

R. Braden, "Requirements for Internet Hosts -- Application and Support, RFC 1123," October 1989.

J. B. Postel, "Transmission Control Protocol, RFC 793," September 1981.

J. Postel and J. Reynolds, "Telnet Protocol Specification, RFC 854," May 1983.

J. Postel and J. Reynolds, "File Transfer Protocol (FTP), RFC 959," October 1985.

S. Deering, R. Hinden, "Internet Protocol, Version 6 (IPv6) Specification, RFC 2640," December 1998.

G. Montenegro, S. Dawkins, M. Kojo, N. Vaidya, "Long Thin Networks, RFC 2757," January 2000.

# 6. Email: The Unforeseen Catalyst

1    J.C.R. Licklider, Albert Vezza, "Applications of Information Networks," Proceedings of the IEEE, 66(11), November 1978.

2    J. Postel, "Summary of Computer Mail Services Meeting Held at BBN on 10 January 1979, RFC-808," March 1982.

3    J. Klensin, "Simple Mail Transfer Protocol, RFC-2821," April 2001.

General Sources
J. Meyers. M. Rose, "Post Office Protocol — Version 3, RFC-1939," May 1996.

M. Crispin, "Internet Message Access Protocol — Version 4, RFC-2060," December 1996.

N. Freed, N. Borenstein, "MIME Part 1: Format of Internet Message Bodies, RFC-2045," November 1996.

N. Freed, N. Borenstein, "MIME Part 2: Media Types, RFC-2046," November 1996.

P. Resnick, "Internet Message Format, RFC-2822," April 2001.

# 7. Cyberspace: The Internet as Meeting Place

1    William Gibson, "Neuromancer," Berkley Publishing Group, 1989, 128.

2    By-Laws of Web Based Chat, from City Live Web Site, http://www.citylive.com/rules.html, July 2003.

General Sources
C. Kalt, "Internet Relay Chat: Architecture, RFC-2810," April 2000.

Dr. Richard Bartle, "Interactive Multi-User Computer Games," MUSE Ltd., British Telecom PLC, December 1990.

Remy Evard, "Collaborative Networked Communication: MUDs as Systems Tools," Proceedings of the Seventh Systems Administration Conference (LISA VII), pages 1-8, November 1993, Monterey, CA.

# 8. Strategies for Success: The Technology and Management of Interoperability

1    M. A. Padlipsky, "A Perspective on the ARPANET Reference Model," RFC-871, September 1982, 17.

2    S. Harris, "The Tao of IETF — A Novice's Guide to the Internet Engineering Task Force," RFC 3160, August 2001.

General Sources
Stephen Crocker, "Making Standards the IETF Way," Standard View, Vol 1, No. 1, 1993.

A. L. Chapin, "The Internet Standards Process," RFC-1310, March 1992.

# A. Milestones, Netiquette, and Jargon

General Sources
Milestone information for the Internet and the Web adapted from the following sources:
Robert H. Zakon, "Hobbes' Internet Timeline," http://www.zakon.org/robert/internet/timeline/, 2003; and Lawrence Roberts, "Internet Chronology," 22 March 1997.

# Index

## B

## C

# D

# E

## G

## J

# N

# S

# U

# X

# Y

# Ironbound Press
# Winter Harbor, Maine

To order copies of this book:

Visit us on the Internet at:
http://www.IronboundPress.com

Or photocopy the order form on the opposite side of this page and send to:

Book Orders
Ironbound Press
P.O. Box 250
Winter Harbor, ME 04693-0250

Or inquire at your local bookstore.

Ironbound Press books may be purchased for educational, business, or sales promotional use at discounted prices. Discounts also apply when ordering 5 or more books. For information, please write to the address listed above, or send email to:
info@IronboundPress.com

# Ironbound Press Book Order Form

Send completed form to:
Book Orders
Ironbound Press, P.O. Box 250, Winter Harbor, ME 04693-0250

| Bill To: | Ship To (if different than Bill To): |
|---|---|
| Name: | Name: |
| Address: | Address: |
| City: | City: |
| State/Zip: | State/Zip: |
| Phone: | Phone: |
| Email: | Email: |

| Qty. | Item | Description | Item Price | Total |
|---|---|---|---|---|
| | 0-9763857-5-9 | The Internet Revolution (paperback) | $22.95 | |
| | 0-9763857-6-7 | The Internet Revolution (hardback) | $26.95 | |
| | | | Sub-total: | |
| | | *Shipping and Handling: | | |
| | | **Sales Tax: | | |
| | | TOTAL: | | |

\* $4.00 for the first book; $2.00 for each additional book.
\*\* Please add 5%, if shipping to a Maine address.

| Payment Method: |
|---|
| □ Visa  □ Mastercard  □ AMEX  □ Discover  □ Check |
| Signature: |
| Name on card (printed): |
| Card number: |
| Card Expiration date (MM/YYYY): |